D0447180

atomic farmgirl

atomic farmgirl

Growing Up Right
in the Wrong Place

Teri Hein

A Mariner Book

Houghton Mifflin Company

NEW YORK · BOSTON

First Mariner Books edition 2003
Copyright © 2000, 2003 by Teri Hein

For information about permission to reproduce selections from
this book, write to Permissions, Houghton Mifflin Company,
215 Park Avenue South, New York, New York 10003.

Visit our Web site: www.houghtonmifflinbooks.com.

Library of Congress Cataloging-in-Publication Data
Hein, Teri.
Atomic farmgirl : growing up right in the wrong place / Teri Hein.
p.   cm.
ISBN 0-618-30241-7
1. Hein, Teri. 2. Radiation victims — Washington (State) — Hanford
Site Region — Biography. 3. Ionizing radiation — Health aspects —
Washington (State) — Hanford Site Region. 4. Nuclear weapons
plants — Health aspects — Washington (State) — Hanford Site Region.
5. Radioactive pollution — Health aspects — Washington (State) —
Hanford Site Region. I. Title: Atomic farm girl. II. Title.
RA569.H445 2003
362.1'969897'009797 — dc21
2002191947

The author would like to thank the publisher for granting permission
to reprint the quote from William Kittredge, *Hole in the Sky: A Memoir*,
courtesy of Alfred A. Knopf, a Division of Random House, Inc.

Previously published in hardcover by Fulcrum Publishing.

Printed in the United States of America

QUM 10 9 8 7 6 5 4 3 2 1

For my family, my neighbors,
and Jim Smith

acknowledgments

Many people have helped me to write this book.

The first was Mrs. Dagley, my high-school journalism teacher, who gave every indication that she found me, in general, annoying and yet still told me she thought I was a good writer. Because I thought she didn't like me, I believed her about the writing. I got my publishing start writing the gossip column "Bugsy's Beat" in our school newspaper, the *Liberty Ledger,* which had a weekly circulation of 150. That column meant a lot to me at the time.

About twenty years later Terry Tempest Williams told me she thought I could write a book. In that two-week workshop, the only formal writing class I'd had since Mrs. Dagley's, Terry inspired me. It is not easy to inspire German Lutherans.

Then there are the people who went to varying amounts of great trouble to help: Charlie Bell, Marlene Blessing, Mary Bond, Mary Bruno, Susie Dixon, Linda Gunnarson, Kitty Harmon, Jackie Kann, Katherine Koberg, Eric Levensky, Rick Loeffler, Bill McCauley, Jim Smith, and Kate Thompson. I live in fear of having forgotten someone.

And, of course, mostly, I thank my parents and my neighbors, who answered all the questions, even the hard ones that made things sad all over again. And my sisters, who have allowed me to embarrass them publicly for my own personal gain.

May this lighthearted thank-you convey my serious appreciation.

preface

My dad is what I would call a survivor of the Cold War. His battlefields were the rolling hills of the Palouse of eastern Washington, one hundred miles northeast of the Hanford nuclear plant. I believe my father uses a wheelchair today because of the Hanford nuclear plant, the quality of his adult life having been significantly affected by the plant's radioactive releases, which invisibly littered the land during the Cold War. I don't know for sure that those releases are to blame—we will never know for sure—but I think so.

What I do know for sure is that iodine-131, a radioactive element, causes cancer, as do a host of other radioactive elements that were strewn over eastern Washington in the 1940s and '50s. Hanford repeatedly released those elements at many times the levels known to cause cancer.

Except for two years in the army, my father has lived his entire life along the wind pathways that stretch from southeast to northeast Washington State and beyond. He was diagnosed with thyroid cancer in 1951 at the age of thirty-one. To save his life, the doctors removed not only his thyroid gland but also parts of his jugular vein. After the surgery he developed a peculiar limp that has worsened over the years. The doctors can only guess that his circulation was irreparably damaged by the removal of his jugular vein, which has affected his ability to walk. His thyroid surgery may also account for the brain hemorrhage he suffered fifteen years later.

I don't care that expensive scientific studies claim there is no definitive proof that Hanford caused cancers. I don't care about judges and lawyers who ping-pong evidence in and out of court—valid one day and not admissible the next—as if it were a game of table tennis. I just care about my father, and he is only one of thousands of people probably, and quite tragically, affected by activities during the Cold War. I want to make sure that intentional releases of radioactive materials never happen again.

I wrote this book over a span of more than ten years, mostly during the 1990s, slowly adding layers to what I already knew from growing up in my two-square-mile neighborhood in the Palouse of eastern Washington. The layers came from reading newspaper accounts and declassified documents, interviewing neighbors, poring over letters from my ancestors, and scratching the memories of my family members.

Atomic Farmgirl was published in 2000, and only then did I realize how much more I had to learn. Only then did I begin to think of my family and our neighbors as veterans in a different kind of war.

"You did the country a service in telling your story. Your people served as much as those who landed at Normandy or Tarawa or Okinawa," wrote Warren Haussler, a physicist who worked at Hanford in the 1950s. He was not with the Chemical Separation Plants and was not aware of the radioactive releases at the time, but a few years later, when he worked at the Atomic Energy Commission in Washington, D.C., Mr. Haussler became aware of what was happening and protested, to no avail.

I met Jim Thomas, a man who for twenty years has worked to unravel the story of Hanford. As a citizen activist in Spokane, he became involved initially through the Hanford Education Action League, becoming one of the first people to examine the 19,000 pages released by Hanford in 1986 through the Freedom of Information Act. Now Jim works as a paralegal on one of the lawsuits and continues to comb through the documents. He told me that *Atomic Farmgirl* is essentially accurate except for one big mistake. I implied that the releases were mostly accidental, the result of faulty filters or bad management. That isn't true. Most of the releases were the result of routine operations and were done at a time when those authorizing the releases were well aware of the health risks to everyone living in the region.

Jim Thomas is a careful thinker, a man who has repeatedly cautioned me against overreaction and unfair assumptions. He also has worked with Hanford data, continuously and passionately, for longer than anyone else I've met. Before I knew him I wanted to believe that most of the releases that so poisoned our environment were accidental. Maybe I elected to believe that because it was easier to forgive human error, faulty filters, and new technology. But when Jim Thomas, who has read the government documents, told me that most of the releases were intentional, I believed him, and my forgiveness turned to anger. Hundreds of thousands of people were living in the pathway of this airborne radioactivity, and the people in charge at Hanford still released it.

In the summer of 2002 I corresponded with Al Conklin, the manager of the Air Emissions and Defense Waste Section of the Division of Radiation Protection for the Washington State Department of Health. I asked him, if the same radioactive releases that occurred in the forties and fifties were to occur today, what his office's response would be. I specifically named the Green Run of 1949, the largest single iodine-131 release at 8,000 curies, as an example.

To put a statistic like 8,000 curies in perspective, think about this: between 1944 and 1972 Hanford released an estimated 740,000 curies of iodine-131, of which the Green Run contributed 8,000. By contrast, it is estimated that the 1979 release from the Three Mile Island nuclear power plant was about 15 curies. Thus the Green Run intentionally released more than 500 times as much iodine-131 as the Three Mile Island accident.

In that context it is not so difficult to understand Al Conklin's response. If something like the Green Run occurred today, he wrote, his office would immediately order the evacuation of most of eastern Washington and parts of Oregon. Any milk products from pastured cows would be destroyed, along with any other exposed agricultural products.

However, Mr. Conklin pointed out, although the largest in terms of curies, the Green Run was not the most dangerous of the releases because it occurred in the winter, when crops were not growing and milk cows were not grazing on open fields. There were many more dangerous releases during the forties and fifties, releases that covered a broader area and occurred during the growing season. Mr. Conklin could only speculate on "the nightmare of responding over several state borders."

Imagine it. Eastern Washington alone covers roughly 37,000 square miles. In 1950 the population was about 650,000 people. Now add "parts of Oregon and several other state borders." How would it have been possible to evacuate these people and destroy the very crops that were the livelihood of so many of them? And how many times would people have had to be evacuated, as repeated radioactive releases fell over the land? It is this kind of information that propels me to believe that my father in his wheelchair is a Cold War survivor.

But of course, Mr. Conklin went on to write, his office, an important part of the Division of Radiation Protection for Washington State, would never have known about these releases, because they were so highly classified at the time. If his office wouldn't have known about the releases, I suppose it shouldn't be so surprising that those of us living under them didn't know about them. In fact, we didn't know about them until they were long past—forty years past in some cases.

Mr. Conklin believes that releases like those we experienced could never occur today. I am not so optimistic. Since writing the book I have read the report of the Advisory Committee on Human Radiation Experiments (ACHRE). This report, released in 1995, is many chapters long. I was particularly interested in Chapter 11, "Intentional Releases: Lifting the Veil of Secrecy," which says:

> Could this happen again? Could there be another Green Run? The answer is a qualified yes.
>
> In fact, an intentional release like the Green Run probably would not be contemplated (because the scientific and strategic value would seem minimal), but actions that raise similar concerns if undertaken in secrecy could still happen. Environmental regulations apply to secret programs, but the oversight procedures are not fully in place to ensure adherence to these regulations. The public review process that is at the heart of current environmental protections could be limited or rendered nonexistent if the government were to invoke exceptions for "national security interest" to avoid these constraints.

Maybe the Hanford nuclear plant will never again intentionally release hundreds of thousands of curies of radioactive material over the landscape, poisoning everyone and everything below, but I am pessimistic enough to think that equally destructive and secretive activities could be justified by the government of the day.

And it is those activities and the people who make the decisions to authorize them that change neighborhoods and change lives and change our sense of security. And isn't that, really and ultimately, what it should be all about—our lives and our neighborhoods?

Here is a book about my neighborhood and how it changed, not just because of Hanford but also because of Colonel Wright and homesteaders and economics and the forces of life that have made all of us, in different ways, victims of wars. *Atomic Farmgirl* is about choosing our battles and remembering what is most important in our lives.

We tell stories to talk out the trouble in our lives, trouble otherwise so often unspeakable. It is one of our main ways of making our lives sensible. Trying to live without stories can make us crazy. They help us recognize what we believe to be most valuable in the world, and help us identify what we hold demonic.

—William Kittredge,
Hole in the Sky: A Memoir (1992)

Gypsy, our Welsh mare, seemed as tall as a house and as wild as the stallion she wasn't when she remembered the clover on the north side of the house and took off. She forgot about me on top, as we loped under a low branch of the hawthorn bush. I grabbed onto the overhang and stayed there, as Gypsy continued on. She left me to dangle for an instant before I crashed down amid cries and giant scratches and fleeting hard feelings about clover and horses in general. I wonder whatever happened to her.

Come to think of it, I don't know what happened to any of them, except Rockette, who died in the back pasture one summer. Gypsy was just one of many horses in our lives, a list that started with Smoky and Patches. Smoky belonged to Cheryl, my older sister by a year. At that time, when I was seven, she was tall (to me), lithe, and almost able to handle Smoky, who was only slightly taller than Patches and only slightly friendlier. Patches, a pint-size Shetland as ornery as tradition dictates the breed should be, belonged to me. Perhaps they suffer from a Napoleonic complex, those Shetlands. After a few years Dad sent the two away. But I still have, Scotch-taped to a piece of steno paper, chunks of each of their manes—long, black hanks of bristly hair, clipped off on their

last day with us before somebody's trailer took them away. They
came together and they left together. That much I do remember. I
felt bad when Patches left, even though he had bitten me countless
times and had bucked me off nearly as many, and one time had
reared up so high when Dad was yelling at him that the horse lost
his balance and fell over, landing on top of Dad. I raced into the
house in hysterics to announce to Mom and my grandparents that
Dad was now dead. How could anyone survive being landed on by
a horse? At seven years old, a Napoleonic Shetland was every bit a
horse to me.

Besides, Dad wasn't a large man, something I knew, even at
seven. He wasn't as tall as Dick Dennie, our neighbor, who was
over six feet. Nor was he football-big, like Carl Groth, Susie's dad,
who looked like a cheerful bulldog and still does. My dad was just
medium in most ways, but not all. He already had his limp,
although I don't remember ever thinking that he walked funny. In
fact, in spite of my fear that a falling horse could squash him like
roadkill, I assumed he could survive almost anything. Maybe it
was his skin. A lifetime of sun on our eastern Washington wheat
farm had left him with a reptilian hide that seemed a protection
of sorts. Dad's hands were always covered with nicks and cuts
from machinery repairs, but he never seemed to bleed much.
Except, of course, the time he almost cut off his arm with the
Skilsaw and left a blood trail from the shed to the garage to the
living room. Dad kicked open the back door with his foot and
announced in his customary understated way, "Dolores, I think I
need a Band-Aid," as his blood dripped onto the linoleum floor.
She just chanted his name—"Ralph, Ralph, Ralph"—over and
over as she wrapped a towel tighter and tighter around his arm
before calling the doctor.

For most of his life, at least the farm part, Dad hasn't had a
visible ounce of fat on him, so hard is the work of managing a
thousand-acre wheat farm, sixty-odd beef cattle, a miscellany of

other animals, and four daughters who, at best, were sporadic help. We were given horses to improve our reliability. We had to keep the tack in order, keep the rocks out of their hooves, wipe them down with fly repellent, use the curry comb ruthlessly, especially in the spring when their winter hair was coming out in thick, airy chunks, and braid their manes and tails for special occasions like the county fair and the Flag Day parade.

Dick Dennie taught us the parts of the horse. Thanks to him, we could easily identify our animals' forelocks, muzzles, cannons, pasterns, fetlocks, barrels, gaskins, croups, and withers. He raised Appaloosas and was the leader of the Cayuse Kings and Queens, our 4-H club. He taught us the parts of the saddle, such as the fork, the cantle, the cinch, and the stirrup. He taught us our club song, "Don't Fence Me In," and how to give demonstrations with poster boards and pointers.

I chose to give my demonstration on the quarter horse. On a large piece of pink poster board I traced a side view of a particularly good quarter horse from a picture I found in our *Encyclopaedia Britannica*. As I talked about the animal, announcing to my fellow club members the horse's most distinctive features, I tapped solemnly with Dick's pointer on the convenient body parts, although not necessarily tapping on the ones I was mentioning, as my nervousness took its own course. "The quarter horse is one of America's favorite saddle horses," I said, pointing to the side of the tail. "It stands about fourteen to fifteen hands high," I continued, nervously tapping. "A hand is a way to measure horses. One hand is about four inches. When a horse is measured, it is from the ground to the top of the withers, which is the highest point on the shoulder." I tried to speak with great authority, pretending that I was the only one in the room who knew this information. "The quarter horse has very strong legs," now tapping its head, "which is why it can run at high speeds for short distances." By now I was knocking on its tail. "It was named the quarter horse because in the olden days," tapping on belly, then head, "owners used it to run quarter-mile races." My pointer was like the blind man's cane, swinging widely to find a sidewalk curb. I still remember where the pastern is—it's a part of the foot—and how many hands the average quarter horse stands (about fifteen and a half).

To foster a sense of community service, Dick directed the Cayuse Kings and Queens to meet on horseback one Saturday morning down at the Colonel Wright monument. Dick rode his Appaloosa stallion, Chief Qualchan, the most beautiful horse we'd ever seen. He was so spirited and temperamental that Dick was the only person who could ride him. Appaloosa mares from all over the Inland Empire were brought to mate with him, a rigorous calling that apparently Chief Qualchan was stud enough to accomplish. The owners of those mares paid a large fee for a chance at a foal with the striking blanket of spirals that the stallion wore. From the front he looked just like a black stallion, maybe just a little over fifteen hands tall, but with such a wild look in those eyes that glowed out from that ebony coat that he seemed much larger. And when he turned, no matter how many times you'd seen this horse, that explosion of spots on his back blanket was startling, a wild collection of gray and black circles set on a striking blanket of white horsehair, like an electric table-cloth folded over his body. I don't know who Chief Qualchan's mother was, but his sire was Hopi, a champion gray Appaloosa stallion, completely covered from head to hoof with spots, from his muzzle all the way down his legs. Unlike his son, Hopi had a gentle personality, or at least as gentle as a high-strung, well-bred Appaloosa stallion can have.

We had named our 4-H club after the Cayuse Indians. Dick told us that Qualchan was the last chief of the Cayuse, although later we found out he was really a Yakama. On that Saturday of civic duty, we children raked pine needles while our fathers installed a picnic table by the monument. There people could eat hot dogs and potato chips in exactly the same place where Colonel Wright had ordered Chief Qualchan and a few of his braves to be hung a century before. Years later I asked my mother if it struck her as ironic.

"What?" she wondered.

"Us," I said, "putting a picnic table over the death site of the Indian our club was named for so people could eat in comfort by the monument that honored his murderer."

"Well, now that you mention it . . . ," was all she said.

My mother was not a horsewoman. In fact, I don't recall her ever sitting on a horse. She would lead the Shetlands around with a nervousness only slightly less than that of whichever youngster was precariously clinging to the saddle horn. Mom was what we called a city gal, even though her sum total of city life had been two years at Pacific Lutheran College in Tacoma. She had spent the rest of her life in middle-of-nowhere parishes with her preacher father and family. She was the only woman we knew who had been to college, which meant she knew loads of impractical things, such as the names of rivers in Africa and how to explain photosynthesis to us. When she referred to a book she was reading, she often said the author's name as if it were part of the title. "I'm reading Hemingway's *For Whom the Bell Tolls*," or "I just finished *To the Lighthouse* by Virginia Woolf," she would say to us, her daughters, who didn't ever care about these things until much later. Mom was uncoordinated and had no interest in animals or gardening. It is to her credit that, as a loyal farm wife, she fanned the flame of her daughters' tomboy tendencies, perhaps feeling some vague guilt about providing her farmer husband with nothing but a pack of girls.

We had many horses in those years. Marsha, my oldest sister, had Shauna, with that distinctly sculpted Arabian face and a lovely way of tossing her head. Shauna was skittish and dramatic, as Arabian horses often are, and beautiful when she pranced around the pasture. She was a deep rusty red with a wispy blond mane and tail, both of which Marsha brushed and braided religiously. During the Shauna era, Marsha was reaching adolescence with the accompanying lumpy body, the daily threat of pimple eruption, and the long, up-flipped hair that was fashionable at the time.

Cheryl's horse was Lady Anne, regal in her way of prancing and bobbing her head as she trotted. She was like Cheryl, who seemed sort of regal to me as we grew up—tall, poised, and always in control. Lady Anne was mostly quarter horse, which made her sturdy and trustworthy in personality. She was black with hints of dark brown around her muzzle and hooves.

Rockette was mine, a Welsh mare of mixed breed, and not regal or beautiful but kind and, because of her docility, a family favorite like Blitzen the dog and Snip the cat. She was the smallest of our horses because of the Welsh in her. She was a dirty red brown with a black mane and tail. I rarely braided her mane, but brushed it at least once a week.

Ross was Tracy's leaning old gelding, with a reddish coat, odd roan spots around his eyes, and a bit of a swayed back. Like Lady Anne, he was mostly quarter horse and reliable. That was pretty much his most dynamic quality. Ross was no leader and harbored no aspirations. With our baby sister, Tracy, clinging to the saddle horn in all optimism, he would amble down the road as long as some other horse was in front of him. Without another horse, he wouldn't budge but, rather, became an autistic stone statue who couldn't be beaten into moving a hoof. Get Lady Anne to amble by, though, and he would about-face like the trouper he was.

Then there were the foals—Cheshla and Mejroc and Shamrock, as well as other ones whose names I don't remember. They came and went so quickly, sold through ads in *The Spokesman-Review* to people who promised them good homes.

We girls were always begging for Dad to breed the mares, even though we knew we wouldn't keep the foals, and I'm sure we never made any money from them. We just loved foals, the newer and more wobbly-legged the better.

Our horses weren't purebreds, but often times were some part Appaloosa, and we frequently chose to breed our mares to Appaloosa stallions. We had a thing for Appaloosas, even the watered-down version. This passion for the spotted horse must have been something in our blood—or, more probably, in our dirt.

The Europeans brought the horses, but the Nez Perce changed them. Thanks to the dirt, the grass, and the Grand Ronde Volcano hundreds of thousands of years ago, we have the Columbia Plateau. This is the way my mind, not that of a geologist, wraps around the story.

The Palouse hills in eastern Washington State—where our farm is, where we had our horses, where the Colonel Wright monument still sits, and where the clover for which Gypsy betrayed me grows—are covered with a kind of dirt called loess. Something like 7,500 pounds of this dirt settles in the form of dust on each Palouse acre each year. Our state history textbooks claim it to be the richest topsoil in the world. It seems a grand claim, but one I prefer to believe.

I also prefer to believe the story a Colville Indian told my friend, Tim, when, as a hitchhiker, he jumped into the man's pickup. According to the Indian, thanks to the loess, there is a kind of grass that can grow in the Palouse that is so tough it can stand a winter, pushing its way through the snow for the horses to eat. The Indians in the area—the Cayuse, the Nez Perce, the Yakama, and the Palus—raised horses because, thanks to the grass, they could feed them all winter. They had huge herds that were not only utilitarian, but also as beautiful as the landscape. After years of careful breeding, round and oblong shapes covered the rumps of these animals. Against the snow, their winter coats were a dizzying display of dapples and circles inside circles of white, black, gray, and brown. The Palouse hills were so rich and the grass so abundant that giant herds of these startling animals spread out over the plateau. Eventually somebody gave these spotted horses the name of their Palouse landscape: Appaloosa.

My history textbook doesn't mention the horses, although it does discuss the dirt. The Palouse hills are large dunes that roll out across the landscape. Because of the plateau's high elevation, the hills receive more rainfall than the lower lands to the west. The top layer of the soil is the loess—a foot deep and dark yellow. Under this is a three-foot layer of lighter soil, and under this is seventy-five feet of red Palouse clay. All of it rests on hundreds of feet of undecayed lava rock. On top of it all are the stands of grass that grab at the dust as it flies by.

How did it look before the pioneers came and plowed every-thing up, organizing it into townships and sections and tracts called homesteads? From a copy of ancient papers I got from Washington, D.C., that include my great-grandfathers' homestead claims and the township map, comes this 1873 description of our township by David Clark, the first federal surveyor of our area:

> . . . consists almost totally of rolling hills, some of them high which are covered in excellent bunchgrass. In the South Eastern portion, the soil is generally first rate, along Hangman Creek it is rocky and unfit for cultivation. In the South East portion are numerous springs of excellent water, while Hangman Creek flows in a northerly direction through the Township. . . . The creek is on the bottom of a canyon, with banks 150 to 200 ft. in height, and quite precipitous along Hangman Creek . . .

I like to believe that Mr. Clark, along with his chain men, Andrew Maxwell and Wayman Crow, mound man William Mulligan, and flagman Felix White, after months in this territory, baking in the heat, scrapping with the Indians, and clinging to a certain feverish determination to plot this land, describing each cliff and gully according to its vegetation and dimensions, must have sometimes just looked out over the hills and thought: How beautiful.

A pamphlet called simply "The Washington Territory," written in 1879, describes the countryside like this:

> Man has scarcely dreamed, in his most extravagant fancies, of an ideal country which has not a counterpart in the vicinity of the Spokan. . . . The natural drives of the Spokan plains are probably unequaled in the world. . . . One may be an ardent admirer of a variety of scenery. We would gratify his curiosity. Here he may ascend to a dizzy mountain height, and, from one position, look admiringly down upon hills, valleys and plains, forests, groves and sparsely timbered sections, rich soil and gravelly prairies, pleasant homes and forest fastnesses, the mirror-like lake and the roaring torrent of the wonderful Spokan, trackless glades and level roads, barren, rocky cliffs, and green verdure and blossoming vegetation, unadorned nature, and promising harvests, the beautiful and the grand.

Today a grid covers these poetic hills. Giant squares of wheat, barley, or lentils butt up against each other like quilts, knotted together by farms. The pattern seems planned by some higher power, or maybe the beauty is only a gorgeous example of great luck—luck that the making of food can also be so lovely to look at. Certainly no group of human beings could have organized to create this giant artwork stretching more than 275 square miles called our Palouse—soft in the spring, intense in the summer, stark in the fall, eerie and lonely in the winter.

I try to imagine how the human Chief Qualchan thought of the Palouse hills as he rode to what he thought was the signing of a peace treaty in 1858. Or how Colonel Wright wrote about them to his wife, Amanda, back in Florida, as he moved through the Washington Territory in the 1850s determining how to possess this landscape. Or how my great-grandfather Hans, on my father's side, described them to his wife, Catherina, when he wrote to her in Iowa in 1885 and said, "Take the train." Or how my grandfather Martin, on my mother's side, heard about them. He was in North Dakota when he decided in 1943 to answer the call to Colfax, in the heart of the Palouse, where he would be the Lutheran minister in a time of world war worry. Or how Dick Dennie saw it as he rode home on his Appaloosa stallion, Chief Qualchan, from the monument that day in 1962 when we put up the picnic table. One hundred years before that day, more or less, Colonel Wright had ordered the horse's namesake to be hung by the neck on the very piece of land my great-grandfather Hans later homesteaded. The Indians called the water running through it then the Lahtoo River. Later the name was changed to what we all call it now, Hangman Creek, a name that over the ensuing decades has seemed to become increasingly more appropriate.

It's weird, when you think about it, how people and events are often connected in some way or another; you know, the six degrees of separation and all that. I think it is true. Like that time I was at Gatwick Airport outside London on my way to New York and met those two Italian guys. Our airline was on strike and we had to wait a whole day together to get out. We hit it off and, after reaching New York, ended up renting a car together to drive across America. We had long conversations as we sped through Tennessee and Oklahoma at night. But it wasn't until we had reached California that we figured out that Giancarlo's mother, who lived outside Palermo, was good friends with the daughter of a woman who lived across the street from my grandmother in Spokane, Washington.

This six degrees can also go back and forth on a timeline. What I'm thinking about here is that Colonel Wright hung Qualchan on the land my great-grandfather later homesteaded. My great-grandfather gave that land to my grandmother, his only daughter, who then passed it on to my father, her only son, who then had all these problems that maybe never would have occurred if Colonel Wright hadn't hung Qualchan in the first place. Well, that might be stretching it a little too far, but who would have thought that Qualchan's death, my father's bleeding brain 160 years later, and all the other horrible events that happened in our neighborhood could have anything in common? But they do—at least I think they do, as do most of our neighbors. All of the events were tragic and all were a result of some strange invasion. At least, we in the neighborhood think this. Lots of other people think it, too—people we don't even know.

* * * * *

I had thrown down my *Algebra One* book in disgust, convinced I was a complete loser in the field of mathematics, and gone to bed. About an hour later Mom came upstairs to tell us that she and Dad were going to the hospital because Dad had a terrible headache. They didn't want us coming downstairs to see him. Instead, Cheryl, Tracy, and I gathered in Cheryl's bedroom, the one that faces south and overlooks the driveway, and watched as they hauled Dad away on that December night when his brain began to leak blood.

There was Milt, who doubled as the town ambulance driver and mortician, Dr. Hart, my mother, and, of course, my father, the man of the hour. It was pitch-black outside except for the eerie glow the porch light cast across the driveway. Big floating flakes fell to the ground, seeming to suck up all noise, quieting the atmosphere. Clouds of white breath surrounded the people's heads as they cautiously pushed Dad's stretcher through the crusted snow on the sidewalk. Blitzen, our German shepherd, circled up and down the driveway, drawing giant protective ovals in the snow with his paws. We watched in silence as they lifted the stretcher into the back of the ambulance. Mom, her coat flapping open and her purse dangling under her elbow, crawled in behind Dad. Dr. Hart was already inside, crouched over his patient, who was also his friend on more, or less, auspicious occasions. The doctor had helped birth all of us girls, made house calls when we got the chicken pox, gave us our childhood shots and yearly checkups, and on this night arrived at our house within twenty minutes of receiving the midnight call. Dr. Hart and Milt had driven out pell-mell from Fairfield on the iciest roads winter had to offer.

As the ambulance crept off into the night, we felt fine about giving Dad and Mom over to Dr. Hart's care. The siren and the red twirling ambulance light weren't necessary on this highway that wouldn't see another car for hours.

Mom had said, "Your father has a very bad headache and we're taking him to the hospital." A headache didn't seem too serious, we thought, although considering the way that Milt had wrapped and tied Dad up on that stretcher bed, like a straitjacketed prisoner, it must have been quite a pounder. I don't believe Mom and Dad had ever left us alone overnight before, but Cheryl was a respectable fifteen years old and had taken on her role as the oldest with grace since Marsha's departure for college. "Get to bed," she directed us with authority when we finally saw the ambulance lights fade from the highway behind Uncle Detlef's field, and we obeyed, if only because going to bed helped to still the confusion.

When we awoke in the morning, we were no longer free from adults. There was Grandma Hein, in the kitchen, making pancakes. She had arrived mysteriously in the night, a minor miracle because she hardly ever drove anywhere since Grandpa's death a few years before, much less drove anywhere on icy roads in the middle of the night.

My Grandma Hein was a gifted pancake artist, a talent that went completely unrecognized in her time except by her granddaughters. Using lumpy batter as her medium, she could pour just about any shape we requested and flip it to browned perfection.

She brooded over each cake until the color—closest to raw sienna—was perfect on both sides, and we were each presented with a masterpiece on our breakfast plates. She could pour a cat lying on a rug and a Christmas tree decorated with bulbs. She learned to pour a Space Needle after the Seattle World's Fair, and Mary riding a burro toward Bethlehem, appearing, now that I think about it, not very pregnant with the Baby Jesus. My favorite, the one I most often requested, was Babe Ruth at the plate, his bat held high above his head, ready to clobber the ball that she placed as a browned dot on the edge of my plate. After I had consumed the whole of Babe and his bat, I ate the dot, the final punctuation mark on my breakfast.

On the morning after Mom and Dad went to the hospital, Tracy requested a snowman, Cheryl wanted a hexagon (she had a geometry test that day), and I asked for a pair of sunglasses. We kept it simple. Grandma looked a little tired.

Perhaps it wasn't astonishing how fast word could travel, considering our ten-family party line. If a neighbor's phone rang in the middle of the night, something was wrong; the possibility of danger gave the sin of listening in a temporary reprieve. It was okay that our neighbor, Harriett Brewer, hearing our phone ring into her house that late night, picked it up. She discovered Dr. Hart's wife on the line, wanting to make certain that Mort and the doctor had arrived safely. "Are you okay?" Harriett broke into the conversation, and Mom replied that Grandma was on her way.

The next morning we got on the school bus as usual. We were always the last kids picked up on the route, because our farm hugged the western side of the boundary with the Spangle School District. We expected to be greeted by a full, noisy busload of other party-liners. On this morning, however, utter silence greeted us. My dad had barely warmed the sheets in his intensive care unit at Deaconess Hospital in Spokane, and yet every kid on the school bus knew the ambulance had taken him away the night before. Not one of them knew how to give a word of consolation. Even Lola, our bus driver, was silent as she handed us each a cinnamon roll.

School was pretty normal except that Mr. Slater, the principal, squeezed my shoulder when he saw me in the hall, and Mr. Powell, my algebra teacher, told me it was okay that my homework wasn't done, "you know, with your father and all." Usually he never said anything when my homework wasn't done. I opted not to point out that I hadn't done my algebra homework not because my dad had made an unscheduled exit on a stretcher the night before, but because I hadn't understood algebra in months. This was due, I could have told him, in part, because earlier in the school year Mr. Powell had told me it wasn't important for girls to understand algebra. Why bother? I immediately thought, regarding the onerous task of learning math. From then on, I concentrated my mental efforts previously reserved for algebra on designing routines for the junior varsity cheerleading squad, of which I was a member. Most girls didn't make it as a freshman, so it was particularly important that I concentrate on high-quality cheering. I owed it to future freshmen girls.

Miss Plummer's eyes misted over when I showed up for English class. She was our new, beautiful mystery teacher who had shown up that fall, from where we never knew. What made her most mysterious was that she had a wooden leg (probably really plastic) and only three fingers on one hand, although the other hand was perfectly normal. She never said a word about what had happened, and, of course, her students never had the courage to ask her, although we were desperate to know.

Miss Plummer was tiny, like a China doll, and very short, even counting her poufy, ratted-up hair. She had a limp, of course, which did not stop her from walking back and forth in the front of the classroom, lecturing us on *Les Miserables* and why Jean Valjean was in the pickle he was in. Unlike Mr. Powell, who

told me I couldn't really be expected to do his subject well, Miss
Plummer told everyone in the class that there was no reason each
and every one of us couldn't finish an incredibly long book by
some dead Frenchman who had written it at least a hundred
years before. We struggled and whined through that book, and
then she made us read *Great Expectations*. At least this one had
people in it whose names we could keep straight.

I never told Miss Plummer that I really liked those books,
and never told anyone I used to read ahead of the assignment. I
wasn't the kind of kid who volunteered many answers in class, so
she probably never had a clue that was the year I kissed pulp
fiction good-bye forever.

Anyway, that day at school, when Miss Plummer's eyes
misted over because of my dad's trip to the hospital, for that
nanosecond she seemed to invite me to lean on her a little. As a
German Lutheran teenager, however, I was trained not only to
stand on my own two feet, but also in the art of denial. I just
moved on into class, certain that Dad's headache was probably
gone by then, anyway, and more than likely he was back in the
shed working on the combine.

At the end of that school day, Lola pushed a Pyrex bowl full
of tuna casserole into Cheryl's hands as we got off the bus. By
then, Grandpa and Grandma Keller had arrived from Oregon.
They had packed and were on the highway within an hour of
Mom's call. It took them ten hours to arrive. Mom was still not
home, but she had telephoned. "Your father is very sick. They
aren't sure if he is going to live," Grandma Keller told us bluntly.
We counted on her for this sort of honesty. To an outsider she
may have appeared a tad cold, even uncaring, but I don't think
anyone in the family ever considered her that way. She didn't
invite us to ask questions about this new information, and we
didn't. The concept of a dead father was too abstract to really
worry about.

Mom was gone for two days that time. Then, one afternoon when we arrived home from school, this time with a chicken casserole from Dona Hahner, there was Mom, sitting at the kitchen table. Dad was doing better, although, she pointed out, he wouldn't be greasing the combine anytime soon.

My father had had a brain hemorrhage. While the neurologists didn't know it at the time, they now suspect that it had to do with the remodel job the surgeons did on his throat fifteen years earlier when he was diagnosed with thyroid cancer. By taking out parts of his jugular vein, his blood was on its own to find new ways to go up and down his neck. Somehow, his body created new pathways in his head for the blood to travel through. This, combined with a congenitally weak area in Dad's brain, caused his cerebral meltdown. I don't really understand it, but one pathway went haywire in a major way on that December night, and Dad ended up in the Deaconess Hospital intensive care unit, knocking rather insistently on heaven's door.

Even in the midst of the worst, Dad was still first and foremost a farmer. After only two days in intensive care, with just about every drug the FDA had ever approved coursing through his veins, Dad decided it was time to get out of bed and go home to feed the cows. In his own loopy narcotic state he struggled to get out of bed. No amount of assurances could convince him that his daughters and Leonard Zehm were keeping the cows in hay. The nurses propped up rails on Dad's bed and still he thrashed. My mother stood guard in his room, soothing him in a low voice, and still he thrashed. Finally, the nurses took off his pajamas. In Dad's nude, drugged state he still wanted to feed the cows, but he knew he would have to get out of bed to do so. While not exactly cognizant of the fact that he was in a hospital, he did seem to know he wasn't alone and that he shouldn't be parading around in the raw. Dad never enjoyed offending people, always a gentleman under duress, even when underdressed.

That was a horrible December 1967, the year of Dad's hemorrhage. Mom took a leave from her teaching job and drove forty-five miles every day to camp out in Dad's room. The roads were terrible, thick with ice, and snow often filled the air.

Only once did she take us to visit. It was a different era, and children, even teenagers, had no place in a hospital. We dressed as if we were going to a funeral, and Mom coached us while we drove up. "He's really sick, and you need to speak softly. Don't cry. He looks bad, but he's going to be okay. Now give him your hand when you go in and squeeze it so he's sure it's you."

Mom shepherded us through the halls of Deaconess Hospital to a distant elevator. It opened onto a picture-less beige hallway with squeaky white linoleum on the floor. Around the corner was the intensive care unit, where a nurse's station was surrounded on three sides by beds, each partitioned off from the next with a flimsy beige curtain. I remember the room being filled with moaning patients and wailing, weeping family members, but that may just be a by-product of too much television in my life.

Dad had the only single room in the unit, a room built off the nurses' station with a giant window giving the ailing occupant a commanding view of the misery occurring outside his tiny room, not to mention a close-up look at the nurses at work. A small waiting area was behind Dad's room.

We had to visit him one at a time, except Tracy, who, at eight years old, was too young to go alone into his room. She went with Mom first, returned after a few minutes, and said, "It's really dark," before clamming up. Marsha, who had come home from college for an early Christmas break, was sobbing when she emerged from Dad's room. Marsha cried at just about every-thing, though, and we rarely took her tears seriously. Cheryl emerged from the room pale and somber. Without a word, she perched herself on the edge of the waiting-room couch and stared at the linoleum. It was my turn next. Thanks to Cheryl, I was terrified. She was the only sister I took seriously, and her pale demeanor concerned me greatly.

The room was dark except for a small light by Dad's bed. He was curled up on his side under the covers, holding his head in both hands, facing the door as I came in. There were tubes everywhere, needles poking into him, and pumps clicking and wheezing. I could barely hear him when he said, "Hi, Teri. I bet you didn't think you'd see your old Dad like this." The words seemed pumped and wheezed out of this feeble replica of my father by those very machines at his bedside. It was as if all his energy was used up just to say that much. After, he fell silent, his breathing heavy, with a rasping crackle coming from deep within his throat. I waited for him to continue about any subject, but he just kept on with the crackling breathing while I shifted from foot to foot. I endured about three minutes of silence and then said "Okay-bye-Dad," as if it were one word, and left. Only later did I remember that I forgot to squeeze his hand.

Grandma Keller took over the household while Grandpa read his Bible, comforted Mom, and visited sick old people who had been his parishioners years before. Grandma Hein was kept on reserve for emergencies, although her presence might have been a little easier on us. We were used to her because she lived closer. She had cared for us often during our childhood, and we'd grown accustomed to her way of doting on us.

Grandma Keller, on the other hand, insisted that we make our beds and eat everything on our plates. Instead of spoiling us into becoming soft, cuddly girls, she thought it best that we buck up and thank God the situation wasn't worse than it already was. She liked to point out that we could have been born to a family in Madagascar or New Guinea, places we couldn't have found on a map for money, but we knew must be the most horrible places on Earth. Great-uncle Andy Paul and his wife, Frieda, had been missionaries to New Guinea, a place so awful that their son, Andy Jr., had lived with us for the whole year his parents were gone. It was hard to imagine places so horrible that children couldn't even live there.

We girls had to get up early enough to feed all the livestock before we went to school. I swear that Grandma Keller, who never seemed to sleep, took a certain pleasure in waking us well before dawn. On those icy, dark mornings, wrapped in several layers of wool, we crunched our way to the barn through the pitch blackness and new snow, the cold stinging our eyeballs as we blinked away the flakes. We had to kick and tug to pry the barn door open after a night of being frozen shut. Bellowing hungrily from inside, the cows urged us on. It took two of us to pull a single bale down the barn aisle, each grabbing one side of the bale by its twine and heaving forward on the count of three. Sleepily, we cut open bale after bale and threw the hay haphazardly into the mangers. The steers got buckets of oats, as did the horses, thrown sloppily across their hay. Before opening the grain-room doors, we had to bang on each to warn the mice of our imminent entry. The cats and the dogs got their rations, thrown in bowls and old truck hubcaps, or right onto the floor if we couldn't immediately find the bowl. We trudged through the cold, from barn to lean-to to the old chicken coop where the horses were, feeding the animals until finally we could retreat into the house, the sun a red rim on the horizon.

By then we were frozen but awake. Grandma would have breakfast waiting for us, usually a pot of oatmeal with raisins, to be eaten before we dressed for school. Mom sat at the table, nursing a cup of coffee and looking distant. In the evenings, Leonard, our neighbor, came over and helped with the chores. We still had to feed the animals, but he cleaned the barns, pulled down bales from the loft, made small repairs, and generally kept things, and us, propped up. Mona, his wife, had had cancer surgery herself some months before, so Leonard was particularly practiced in the art of propping up.

Once it became clear that Dad was not going to die, Grandpa and Grandma went back to Oregon, and, under the direction of Lola, our bus driver, the party line took over. Every night as we were getting off the bus, Lola handed us something for supper. It could have been Dona's chicken divan, or Barbara's meatballs and a coffee cake made by Claudia. Whatever it was, we devoured it that night and returned the dish in the morning. Lola always remembered whose dish was whose, handing it to the appropriate child as he or she was getting off the bus, and reminding the next day's casserole kids to tell their mothers. Every Sunday, Mom had Pastor Mueller print in the church bulletin a word of thanks to everyone for their help and prayers. We didn't miss church once during this time. The pastor always asked Mom to stand up and give an update on Dad's condition, and then the congregation would join in a special prayer for Dad. We were praying up a storm in those days and needed all the help we could get.

It was almost Christmas when Dad got out of the hospital—Christmas Eve, to be exact. My mother brought him home with the assistance of Aunt Burnice, who met them at the hospital in Spokane and followed them home in her car. The two women, one on each side, helped Dad into the house. Just as we had watched him leave, strapped onto the stretcher, almost one month before, we daughters parked ourselves by the upstairs window to watch Dad's reentry into the house, this time on his own two wobbly feet.

It was disconcerting how very weak, thin, and old he looked as he shuffled into the house. Dad was forty-seven years old. He made me nervous, but not half as nervous as he made my mother, who fluttered around, demanding we keep our voices down, asked him every five minutes if he needed anything, and generally tried desperately to protect him from any further calamities.

The home movies of our mini-Christmas parade stopped that year. Thanks to Dad, we are lucky enough to have home movies that chronicle almost every important event in our young lives, indoors or out—my father was an all-season moviemaker. On Christmas mornings, Mom lined us up according to age on the upstairs steps, where we waited impatiently while Dad set up his lights in the living room. He focused them on the tree and the Santa presents left underneath the night before. Finally he said, "Ready," Mom ceremoniously opened the hallway door, and we emerged. Tracy was first, the youngest, in her footed pajamas with the Plutos printed on them. She was a darling, chubby child. When she was very young, her curly, blond, wispy hair ringleted down her back à la Shirley Temple, giving her an angelic appearance that wasn't always played out in her personality. I was next, sporting a pair of flannel pajamas with Hopalong Cassidys on horseback galloping across my body. Cheryl, behind me, generally went for a solid-colored long nightgown, and Marsha, the oldest, would be wearing her blue baby-dolls, which for anyone else could have been, with the first breath of fall, a certain call for pneumonia. For some reason, Marsha could sleep in those baby-dolls year-round, even with her window open a crack. "For the fresh air," she would tell Mom and Dad, who protested the loss of heat. Marsha wanted to be a singer someday and was often concerned with the care of her voice, a care that apparently included the intake of substantial quantities of cold air.

But on this, our Brain Hemorrhage Christmas, Dad was no more capable of holding up the camera for the duration of our parade than he was of single-handedly filming *Gone With the Wind*. His biggest accomplishment that day was walking with assistance from his bed to the white chair by the Christmas tree. His other accomplishment was making it to the dinner table and eating four or five bites with all the gathered relatives before he went back to bed. He was there with us when we said grace, giving thanks for many things, the passable winter roads, the Christmas gifts, our family together, and, mostly, for Dad's presence at the head of the table.

I don't remember much about Dad's healing except that he had all winter to do it. We continued to make the crack-of-dawn feeding time out at the barn. Leonard still came over and repaired things. Gradually Dad began getting out of the house, riding in the passenger seat to church, while Mom crept over the winter roads. He had rarely ridden with her before and couldn't believe what a wimp she was behind the wheel, her timidity exacerbated by the fear that her husband's brain could start hemorrhaging again at any moment. The memory of that night in hell, just a few months prior, was still quite vivid.

Then one day Cheryl and I were outside school waiting to be picked up by Mom and Dad for a trip to Spokane. When they pulled up we saw that Dad was driving. "Fasten your safety belts," was all he said as we settled ourselves tenuously in the back seat. Mom cast a look of guarded optimism our way, however unconvincing. She gripped the seat as Dad gunned the engine. We made it to Spokane and back without incident, and from then on Dad was, once again, in the driver's seat. That was the day we stopped regarding him as sick and feeble and began to think of him as just Dad again.

For decades, our telephone party line had served as information central for the residents of Hangman Creek. That's what our neighborhood is called, a roughly two- or three-square-mile chunk of land homesteaded and lived on by more or less the same families for more than a hundred years. When Dad was a kid there were ten families on the line. When I was a kid it was the same number and roughly the same families, just a generation later.

In the early days, you could summon any family on the line with the ringer button. The phone button pushed in one house could be heard in every house on the party line. You just had to know each family's code ring to specify which family you wanted to pick up. My grandparents were two quicks, the Lemons were a long and a short, Uncle Detlef and Aunt Mary were three quicks, the Zehms were two longs, and so on. Everyone knew everyone else's code and, hence, always knew who was getting a call. When anybody got a phone call, everybody got a phone call.

Phone conversations were a little bit like postcards: open to everyone. The phones rang, and many people picked them up. Only one person said hello, and the others listened quietly.

Everybody knew that others listened. In a way, it was simply being neighborly.

Briefly during the Depression, my grandparents were without a phone. They couldn't afford it, plain and simple. Grandpa disconnected the brown telephone box and it hung uselessly on the wall. My grandmother's sadness about losing daily contact with her neighbors propelled Grandpa and Uncle Detlef to hatch a telephone plan.

Uncle Detlef Jons and Aunt Mary lived over the hill. Mary was Grandpa's sister. They had a few more resources than most and managed to keep their phone hooked into "Central" throughout the Depression. Grandpa and Detlef ran a line from Grandma's phone to the barbed-wire fence that ran along the western side of the house on the other side of the road. This fence continued over the hill to behind Detlef and Mary's house. Then the men connected a line from that fence to Detlef and Mary's phone. Whenever Grandma wanted to talk, she just had to push the ringer bell and summon Mary. This new setup was not perfect. It didn't allow Grandpa and Grandma to be on the party line, with its much-dwindled list of participants, but it did allow them to call Detlef and Mary, albeit with a scratchy connection that often denied a good, lengthy conversation. This connection smoothed over a few rough edges of Depression isolation, as well as put Mary in charge of any calling Grandma or Grandpa needed to do, since she could still call out into the world.

Mary and Detlef continued living on their farm well into their old age, long after Detlef had retired from farming, and long after my grandparents had moved to the city. I remember them both clearly. He was a big man, jovial and completely deaf. He shouted everything he said, and laughed in a thunderous way that delighted us as children, mainly because it made the buckles on his overalls clatter like jingle bells. Aunt Mary was gentle and kind, and had a million tiny, tight spit curls covering her head.

She made fabulous sugar cookies. To this day, whenever we try to duplicate her recipe, we call our lame attempts "Aunt Mary's." I don't remember her as pretty, but photos of her as a young woman show a stunning person with thick dark hair, heavy-lidded languid eyes, and olive skin, if the tinted old photos aren't too deceptive.

We liked Uncle Detlef and Aunt Mary a great deal and were sad when they died, one right after the other in the same year, when I was eight or nine years old. Mary died first, although Detlef had been sick much longer. He had bone cancer and made several trips to the Mayo Clinic in Minnesota in search of a cure, which remained quite illusive for a very long time. In fact, he had thirteen different surgeries, as the doctors there chipped away at his cancer. After one particular visit the doctors at the Mayo Clinic sent Detlef home with a prescription, asking that he have a local doctor administer it in shot form. Without a doctor handy, and not one to make a big deal about anything, Detlef asked the Spangle vet to come over and shoot him up with a syringe full of this mystery medicine. Only after a week did Detlef learn from the Mayo Clinic that there were several doses in that prescription, meant to be divvied up over several weeks, not in the one giant injection given to Detlef by the vet. Uncle Detlef became a topic of medical journals nobody around Hangman Creek ever read, since his cancer, after that gigantic injection, effectively never returned. He remained cancer-free for the rest of his life, which was several years, and never had another surgery. A big, sturdy man, the only thing that seemed capable of killing him was the death of his wife, since he went a few short weeks after her, just up and out of the blue.

The Jons farm was sold to Gayle Suksdorf, a young farmer from a family down the road toward Spangle. We knew the Suksdorfs, and liked Gayle, so we let him easily into our Hangman Creek community.

After the Depression people had their phones hooked up again and neighborhood networking was back in business. My father went off to fight in World War II, as did many of the young men on the party line. Mothers read letters received from foreign countries over the phone line to any neighbors who picked up on that ring. It was like having access to a kind of Voice of America, these reports from the front lines delivered in the voice of Marie Hein or Pauline Rohwer or Christine Rasmussen. Everyone always hoped one of the boys would call from Europe, but none ever did. The soldiers weren't allowed, unless they won the special drawings the officers held periodically with first prize being a call home. Unfortunately, this piece of luck eluded the Hangman Creek boys.

My father was late in going to the war. In fact, he was with troops on a ship headed toward Germany when the war ended. They had cautiously zigzagged across the Atlantic, hoping to avoid an attack, only to arrive in Scotland to find that the war was over. Dad told stories about being in postwar Germany, traveling to his post in the eastern part of the devastated country over makeshift pontoon bridges through piles of rubble that represented former cities. Later he tried to describe to me the feeling of fear in the air. He said it hung there like the smell of rotten milk or drying manure. That was about as poetic as Dad ever got.

His company was posted outside Berlin, where they shared duties with a company of Russian soldiers. Dad enjoyed the Russians, who often came over to their camp to visit in whatever combination of languages they could manage between them. When the eastern part of Germany was given over to the Russians, Dad's company left, watching behind them as the Russians rolled in their oxen and carts to replace the Americans' tanks and trucks.

Dad returned to the States and called from Mississippi with the news that he was coming home. There were so many neighbors rubbernecking, as we called it, on the party line that my poor grandmother couldn't make out the details over the fluttering and the static (listening in somehow weakened the telephone connection). It was only after Dad hung up that Aunt Mary told Grandma that he was coming home the next week and that he would be flying to Geiger Field outside of Spokane: "Ralph said that maybe they'll circle the farm in their airplane!"

When he begged the favor to buzz the farm, my dad had no idea how hard it would be to find from the air. There he was, inside the nose cone of a B-17 with the pilots while four other soldiers lounged in the back. The plane, considered the best bomber ever built, was doing just what it was famous for— bringing 'em back alive. They had flown all the way from Mississippi. In spite of the soldiers' eagerness to get to their own homes, they were willing to swoop around the countryside, eyes scouring farms below, looking for Ralph's place. Their plane, motoring above the fields, was a sight and sound to experience on the Palouse landscape. These boys were back from the war and they didn't mind everyone knowing it.

As the first drone became audible, the party line went to work. My grandmother shouted into the mouthpiece, "Wave your dishtowels!" The women bolted out of their houses, waving large sheets of white cotton back and forth, to welcome Dad home. What he saw from 1,000 feet in the air was the quilt we call the Palouse, its rolling wheat fields separated by green alfalfa draws and dotted with farms. On each of these farms was a white dot. As the plane swooped down close, Dad identified the dots as dishtowels flapping from the hands of the neighbor women. It was when Dad saw the towel-flags that he knew, after two very long years away, that he was really home. In the fields between the farm buildings and the towels, the men stopped their plowing and slowly moved their arms back and forth from on top of their tractors. "Welcome home!" they yelled up to the airplane.

My grandmother, a tall woman, strong and calm by nature, furiously waved her cloth. "HEL-LOOO!!!" my father yelled at the ground below, the long "O" blowing out of the crack in the window and spreading across the hills. The pilot pushed in the yoke, lowered the nose, and dove down to within seventy-five feet of the ground, circling the farm precariously with a tipped wing so Grandma could see Dad's face pressed against the window. Then the plane shot up, leveled off, and headed north. By the time it had disappeared, my grandfather was warming up the car to head for Geiger Field.

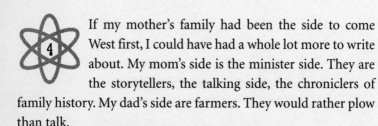If my mother's family had been the side to come West first, I could have had a whole lot more to write about. My mom's side is the minister side. They are the storytellers, the talking side, the chroniclers of family history. My dad's side are farmers. They would rather plow than talk.

When Abe Lincoln decided to encourage a westward movement with a land giveaway, all four of my paternal great-grandparents hit the westward trail. The Heins—Gottfriet and Christina—came first, only a few years after David Clark completed his 1873 survey. They undoubtedly arrived in a wagon. The first railroad came out from Puget Sound in 1881, but the first that came across from the east was completed with a golden spike in Helena, Montana, in 1883. The Thams—Hans and Catherina—came in the late 1880s, riding on that railroad, sitting up all the way since they didn't have the funds to rent a whole train car, like some, who could then bring everything from beds to the family cow.

Arriving when Washington was still a territory, both couples claimed a quarter of a section, 160 acres in all, which is what Lincoln's Homestead Act allowed them. Some people then broke their 160s down to 80-acre pieces, thinking that on this soil 80 acres was all a farmer needed and could handle. Much of the Thams' piece was the rocky land around Hangman Creek, and they used it only for grazing livestock. The Heins, who worked all 160 acres, could have been nothing but industrious. While not next-field neighbors, the two families shared the same neighborhood. True "neighborhoods" in our Palouse, then and now, were giant chunks of land that today might support a city. Proximity was measured less in miles and more in shared plight and industry.

It was Grandma's side that took the Hangman Creek piece below Rattlers Run, and also the prairie that stretches up from the river to where the white house that Great-grandpa Hans built still stands. The Lahtoo River was already renamed Hangman Creek when the Thams took possession. Hans let a group of Cayuse Indians continue to live down by the creek, living in teepees right next to the wagon road and not so far from the hanging tree—they themselves sad reminders from another time.

It would be, among other things, quixotic to describe the life of the Indians as perfect before the white settlers came, but it probably wasn't so bad, if you didn't mind a diet heavy in camas root and salmon. Then the Europeans arrived, trouble broke out, and people died on both sides.

Colonel Wright, the General Custer equivalent in the Palouse, was a career military man who spent forty-three years of his life fighting in the armed conflicts of the age—the Mexican War, the Third Seminole War, the Indian Wars, and, finally, the Civil War. He was a true patriot, according to some, but to others, a ruthless, unrepentant man who hung Indians for sport. His aim was not only to win the Indian Wars of the Pacific Northwest, but also to so dispirit the Indians that they would never rise up again.

Wright's tenure in our immediate area was short, epitomized by two pertinent events that happened toward the end of the wars. The first occurred in 1858, when the cavalry, led by

Colonel Wright, came upon a large herd of horses owned by a band of Cayuse Indians near what is now called Liberty Lake. The colonel ordered his soldiers to round up the animals, which numbered nearly nine hundred. After choosing two hundred of the best ones to keep, the soldiers shot the rest, dropping the animals one by one into a giant pile of hate and revenge while the Indians stood watching. The horses were mostly Appaloosas. Wright and his men left seven hundred bodies of those exquisite spotted animals in the meadow by the side of the lake, in full view of a grieving Indian nation that, by that late date in the wars, was powerless to stop anything. The place, located about twenty miles northeast of Fairfield, became known as the Bone Yard. For decades the remains of those horses lay bleaching in the sun, and the area lay unclaimed by the homesteaders, who must have felt the inherent evil lurking there.

But the event that most characterized Colonel Wright's appetite for vengeance was when he lured Qualchan, the chief of the Yakamas, to his camp by the Lahtoo River, or what we call Hangman Creek. The story, with its various versions that have sprouted over the years, is that only a few weeks after the killing of the horses, Wright took Owhi, Qualchan's father, captive, since he thought that Qualchan was responsible for the deaths of some miners in the southern part of the territory. Wright sent word to Qualchan that if he came and signed a peace treaty, he could reclaim his father. Of course Qualchan came, accompanied by his wife, his brother, and a small group of braves. As the story goes, Qualchan's wife, Whiet-alks, the daughter of Saulotkin, the chief of the Spokane Indians, was both strikingly courageous and beautiful. Qualchan's brother, Lo-kout, carried a rifle; Qualchan, a decorated tomahawk and pistol; and Whiet-alks, a lance, which she had wound with colored beads and ribbon. From the end she hung two long tippets of beaver. Their clothing was beaded buckskin with pictures of the sun, birds, and other patterns of their tribe. They wore loops of shells in stiff necklaces that descended the length of their torsos. With grace and confidence the threesome, followed by the braves, rode directly to Colonel Wright's tent and asked in their Chinook language for Owhi. Wright emerged from his tent, immediately ordered his men to capture Qualchan, and there, on the spot, with no more ceremony, ordered that the chief be hung.

Early accounts, the ones that herald Wright as a hero, claim there was a trial before they hung Qualchan. Wright's journal reports, "Qual-chew came to me at 9 o'clock this morning, and at 9 1/4 a.m. he was hung."

Upon seeing what was happening, Qualchan's wife twirled her lance over her head, uttered a shrill cry as she drove it into the ground, and rode off into the Palouse to spread the word that Qualchan was dead and the wars were indeed over.

Thirty years after this event, in 1888, my great-grandfather, Hans Thams, filed his homestead papers, having claimed the Hangman Creek piece for his family. We don't know how much he knew of his land acquisition's history or if the deadly legacy even bothered him. We know only that he came out before the rest of the family, presumably establishing this piece as their homestead site. Catherina and their early children, which included my nine-month-old grandmother, Marie, followed months later.

When they arrived in this part of the Washington Territory, the railroad had run track down to a platform called Regis, which later became the town of Fairfield, ten miles to the northeast of what is now our farm. Having a railroad platform was essential for farmers who wanted to send their wheat out into the world to consumers.

In those days the railroad companies were inventing towns right and left, hoping to lure pioneers to these barely charted lands, where they would start farms and use the railroads to ship their wheat. Lonely platforms in the middle of prairies were given

impressive names such as Ritzville and Grandview and then, in
leaflets that floated around midwestern cities, described as new,
modern towns sitting in the midst of homesteadable land. Or
platforms were given the name of railroad executives or their
friends—for example, Cheney and Pullman.

I wonder who Regis was, how he managed to get a railroad
platform named after him, and if he was alive to experience the
disappointment that would have come when his railroad platform
was renamed "Fairfield." Colonel Morrison, an esteemed member
of the Regis community, renamed the town to appease his wife
who had given up her Illinois town of the same name to come to
this territory. My father told me once that there is a Fairfield in
every state (although maybe not Alaska or Hawaii), a gentle name
leaving its sentimental stamp across an America of homesick
immigrants.

* * * * *

When my grandmother was a child she attended Rattlers
Run School, a one-room schoolhouse on a bank overlooking
Hangman Creek. A handful of children were educated there by
any schoolteacher available. On Fridays they had spelling bees,
often attended by parents and even some Indians who would ride
up on their horses and peer in the windows as the children
spelled their best.

Mostly the children ignored the Indians. By then, the 1890s,
children weren't afraid of them, not in the way they had been ten
years earlier, when plenty of stories were available, true or not, of
Indians kidnapping or terrorizing the settlers' children.

The Indians were defeated not only on the battlefield, but
also in spirit. Chief Joseph had made his last, fruitless attempt to
retain the Wallowas in northeastern Oregon for his Nez Perce
reservation. The whites had sheepishly justified their takeover as

a somewhat unjust means to an end: thanks to them, they imagined, the once-wilderness was now like a Garden of Eden.

The Louies were survivors. This Indian family had an 800-acre piece over on the Coeur d'Alene Reservation, in what had newly become the state of Idaho, where they raised Cayuse ponies to sell. They were an enterprising family who had resignedly factored in the pioneers' presence as an economic watershed. The Louies sold only the Cayuse, a term synonymous with "Indian mutt pony."

Fat Louie was the family traveling salesman, riding the countryside with the occasional barely broken pony or two in tow. His giant body dwarfed his own pony, which plodded along loyally beneath him. My grandmother remembered Fat Louie as uncharacteristically open and friendly, unlike the usually somber Indians. He was popular with the pioneers and even spoke a smattering of German and Danish. The children gathered at the school window when they saw him coming for the spelling bees, both because they liked him and also to watch him dismount. Fat Louie was so hefty that he found it difficult to get on and off his pony, so he had trained his preferred mount to squat down like a camel, allowing him to roll off without fear of injury.

One day Fat was riding his pony along Hangman Creek, down the trail from the schoolhouse, when the animal tripped right by Turtle Pool. It was the deepest part of the creek, a muddy water hole often frequented by box turtles. Unable to regain its step, the pony fell over into the water. Fat must have hit his head on a rock, because there they found him, drowned and dead, a human island in the center of the water, his belly sticking up above the water line.

Grandma Hein went with her family to Fat's funeral. He had done business with every pioneer in the area, and hundreds of people, Indians and whites alike, showed up. From then on, the schoolchildren, who had often taken their recesses at Turtle

Pool, took to calling the place "Fat's Pond" in his honor. After the accident, they tended to avoid the place, thinking too much about Fat's lovely dismounts, thoughts that brought on small waves of sadness.

* * * * *

Hangman Creek played a major role in the sugar-beet industry when, in 1899, Colonel Morrison and his partner, D. C. Corbin, got the railroad to run some tracks from Fairfield down to Waverly, where they put in the sugar-beet plant. Waverly is between our farm and Fairfield. The entrepreneurs needed our Hangman Creek, which curls all over the countryside, including through Waverly, to wash off the beets before shipping them out into the world. It's hard to believe that Hangman Creek was ever mighty enough to compel someone to build anything on its banks, but the colonel, a Civil War hero and local businessman, did just that. I think the creek used to be more like a river than it is now, so vital it is to the history of our area.

At first only three people lived in Waverly to oversee the initial operations. But then the Asians came to work. Some called them Chinese; some called them Japanese; their origin is lost to us now. Whoever they were, these industrious people, driven by desperation, propelled Waverly to be a town of renown, if only for the short-lived reign of the sugar beet. In the June 26, 1899, issue of the *Spokesman Review,* Mr. Corbin is quoted as saying: "We now have more than 2,000 acres set out in beets and are employing about 300 laborers, 100 of whom are Japs [sic]. Some difficulty has been experienced in getting steady, white hands to work. The trouble with most of the white laborers, we have found, is that they are badly afflicted with the 'road disease.' They will work well enough for a few days, then draw their pay and 'hit' the road in search of more flowery fields. The Jap [sic] laborers, on the other hand, are steady and reliable, and in most cases may be counted on to stay the season out."

By 1900 Waverly's population was officially 918, making it nothing short of a city by the standards of the time. There was a hall for revival meetings, eleven saloons, three churches, two stables for livery and stagecoaching to Fairfield, six stores, four hotels—two of them three stories high—harness and blacksmith shops, merchandise stores, a Chinese laundry, a newspaper publisher, restaurants, and wooden sidewalks on all the streets. Sunday baseball games were renowned, with a league pitcher brought up from Salt Lake City.

By 1910 the bottom dropped out of the sugar-beet industry, and the population of Waverly dwindled to 318. Today there are about 100 people and three operating businesses, if you count the post office. There are the fertilizer plant and the tavern, which was recently renamed Hangman Creek Bar and Grill. There are no sidewalks in town, the schoolhouse is boarded up, and the churches are long gone. There isn't one Asian in Waverly, and most of the residents don't know that for a few years in the way-back history of their town, there was a "Chinatown" covering about three city blocks down by where the Rooney place sits.

Leonard Zehm, our neighbor to the north and the man who helped us most when Dad had his brain hemorrhage, was a direct descendant of Colonel Morrison. The colonel was Leonard's great-grandfather on his mother's side. I know that only because one day when his wife, Mona, and my mother were looking at some old photos found in the church basement, they came across one of the Morrison mansion. The photo was from way before the church bought the building and turned it into a nursing home. Mona had been complaining a bit about how small the Zehms' house was for their family of seven. She sighed as she looked at the giant house. "Should have kept that one for ourselves," she said with a smile.

The Zehm family came a little later than most of us to Hangman Creek. Only after my great-uncle Hans Thams Jr. died, and the Thams homestead was for sale, did Ervin, Leonard's dad, actually own something around the creek. Before that he rented a piece of land a half mile west up the road. Hans was my grandmother's brother, an odd loner who never lived away from the family farm even once in his life. When Hans Sr. died, Hans Jr. took care of his mother, Catherina, or, as some say, she took care of him, since he didn't do much at all, except take photographs, his one passion.

No one wants to talk about my great-uncle, although it doesn't seem as though it's because there are terrible secret stories about him but, rather, because he was mostly a benign oddball who lived his adult life alone. That attention to solitude can make people nervous. He lived mostly on farm income, although he didn't do much farming himself, and on handouts from family members. He sold his share of the Hangman Creek property, the part on the other side of the bridge from ours, to the Zehms for spending money. It was Ervin, coming over to make a land payment one afternoon, who found Hans, dead and all bloated up behind the shed. He must have died long before. To this day Leonard won't tell me the details, so grotesque are they and vivid in his mind that he doesn't want to burden the great-niece, who never once met the victim, with a vision even Leonard himself had only heard about—in detail. Grandma never talked about Hans, although I know my grandpa let her give him money now and again.

Leonard was Ervin's oldest son, and somehow the two of them patched together enough properties and absentee landlords to farm enough to live. Leonard was in Japan after World War II and spoke often, always softly, about his work there. He still talks about it. It is as if Leonard has had two lives—the one on the farm, which so far has comprised about seventy years, and the one in Japan, which was for one year.

Leonard's Japanese life seems, in spite of its relatively brief time, to remain as significant to him as his farm life. His work in Japan was to help sort out the mess of post-bomb Hiroshima. He arrived there six weeks after the United States dropped the bomb and has never forgotten the image that place branded in his mind. He spoke of the twisted bridges, still functional, but wrapped in curlicues, like giant pieces of blackened saltwater taffy. Whole buildings, completely intact on one side, were nothing but burnt-out caves on the other. But the worst memory, Leonard said, the one he most wishes he were rid of, was that of the people. "The people in the center of the blast were the lucky ones, from what I saw. It was the ones who didn't die instantly, the ones on the outskirts, who suffered the hell of what we'd done to them. You can't describe people with their faces melting off," he said, "so I won't even try," and then he would go on to describe matter dripping off their heads as people plodded aimlessly down the streets. I was asking him questions one day about those war years, and he finished quietly with, "It's hard to understand how we could do that to other people." Then he walked away as if he'd just remembered an errand he had to do.

Leonard was a tenor in the church choir, and sometimes I wondered, with the way he talked about Hiroshima and all and how he couldn't forget it, if maybe all that loud singing was just a way for him to get out that memory, as well as deal with the other stuff that made his life harder than others. When Mona got cancer, he often wondered why it was she who got sick, and not he, who had walked for months through the steaming leftover radiation of Hiroshima and come out with nothing but some horrific memories of what people can do to each other in the name of things none of us really understand.

So Leonard came home from Hiroshima and married Mona Lundstrom, the girl down the road whom he had known for most of his life. And they didn't move into the Morrison mansion in Fairfield because it was too far from the fields, but instead rented the old Roth place, just down Cahill Road from us. It was a tiny house, so it was not surprising that every now and again Mona might have wished for the Morrison mansion.

I'm sure she didn't wish this most of the time, though. None of us at Hangman Creek would have given anything to live in Fairfield, which is where the Morrison mansion was. We liked our living arrangements just the way they were, on our farms. Even if some of the houses were small and lopsided, the views were big and our neighborhood was connected not just by fields, but by friendship and history. Our ancestors had worked together plowing up the bunchgrass, building a schoolhouse, and clearing for roads. The next generation worked in the sugar-beet factory, threshed the fields together, and line-danced at Fourth of July events down on the Flat. My father's generation shared car trips to swimming lessons at the YWCA in Spokane, formed 4-H clubs together, and helped each other out when people got sick, which started happening at an accelerated rate with my dad's generation.

If the Zehms had lived in Fairfield, Leonard wouldn't have been around to pull bales from the loft for us when Dad's brain leaked. We would have been rescued by Jimmy Hahner or Dick Dennie, most likely. And what about Mona's brain surgery? She got sick six months before Dad's hemorrhage. As she lay in that hospital bed Leonard rented for their house, Mom drove the mile north on Cahill Road to the Zehms' once a day for two months as Mona recovered, just so she would have an adult there while Leonard worked the fields. When he came home at noon, Mom drove back down to our house and put a late lunch on the table for Dad and his hired men. Dona Hahner took the afternoon shift with Mona, having fed her harvest crew at eleven.

Our neighborhood spent a fair amount of time helping one another out during those years—much more than now. Lucky for us, finding help really wasn't a problem then.

That December brain hemorrhage, as it turned out, probably wasn't Dad's first. Once it had occurred, he remembered the other headache. That particular time the trouble didn't really start with a headache, but actually a soreness in his feet, a kind of odd, dull ache that he never understood and that just eventually went away, or maybe just got upstaged by the bigger pains to come. It was the sore feet, though, that took him to the doctor. The first time Dad noticed them he was on what turned out to be his last hunting trip, a couple of years before I was born.

For a while, according to family legend, Dad was quite the duck hunter. He would go out with his friends, usually Bus Sperline and his brothers, piling into their pickups, the shotguns stashed behind the seats. Dad had a Remington 12-gauge that Grandpa had given him as a child. Dad and his buddies could have been poster boys for the National Rifle Association, all growing up with guns and all quite careful, even after only brief instruction from their fathers on the rudimentaries of these machines. No one ever left the gun loaded within reach of a younger sibling, accidentally shot a neighbor, or maliciously sniped at animals, much less flipped out in a crazed shooting spree during school recess. They were just farmboys who learned early how to take care of themselves.

Dad loved hunting ducks. He loved sitting icy cold on the side of a pond at dawn waiting for hours sometimes for the birds to land on the water. He loved scrunching together with friends in the duck blind, whispering silly jokes the way guys do, often at each other's expense, although not in a mean-spirited way. He loved eating those tuna sandwiches prepared by Mom. She wrapped each individually in waxed paper and then tied the pile together with a string. Dad's fingers, icy and stiff from the cold, hampered him as he tried to loosen the knot without smashing the bread in his hungry haste. He could eat four of those sandwiches without a pause to rest, with or without the inch-thick cut of iceberg lettuce Mom put on often for the green and the texture.

Mostly, Dad loved how the birds sounded when they arrived, pushing out that bird *hee-haw* with each respiration, the *whomp-whomp* of their wings as they braked to a jittering stop on the pond. And I suppose Dad must have loved slowly pulling up his shotgun, resting it against his shoulder, swaying the sight back and forth just a tiny bit until it rested right on the bird, and then pulling the trigger.

As a child I saw many pictures of our front-yard fence lined with dead ducks, hanging by their necks from a wire, so many that I didn't want to count them. Dad and his friends stood behind them in the photos, their guns resting cavalierlike on their hips, giving the impression that this was just all in a day's work, a very fun day of work, if those small smiles with jutted chins said anything about these young men.

They often drove over to Idaho to remote places where they wouldn't be caught shooting in the off-season, or just down the road to Uncle Detlef's pond. They gave the dead ducks away because none of the guys wanted to pluck or gut them. They just wanted to shoot them, at a time in history when that wasn't anything to be ashamed of.

It was on one of these trips that Dad, sitting there in that duck blind, sensed the dull ache coming from the bottoms of both feet, an ache that definitely wasn't to be confused with a pulled muscle or a stubbed toe, but a distinctly different, odd feeling that spread out over both his soles.

Marsha was three years old and Cheryl only nine months, and after Dad complained about the ache for one week running, he went to the doctor. While there he said, "By the way, what do you think this lump in my neck could be?" It was at that moment, you could say, that everything small changed into something very big, and things in my family's life were never the same again.

Forty-eight hours from the moment the X-ray machine flashed, my dad was under the scalpel. He was thirty-two years old and had cancer, an illness encapsulated in a word less familiar to people in 1952 than polio or whooping cough. Uncle Detlef, with his bone cancer, was the only one the family knew who had ever had cancer, and also the only one they had ever heard of who had survived cancer. No one in the family had the slightest idea of how to react. Dr. Hart told my mother that no one as young as my father ever got thyroid cancer, a statement that was apparently quite untrue.

Mom dropped my sisters off at Grandma and Grandpa Hein's house in Spokane. Grandpa Hein was down on the farm, repairing a tractor. It was December. Grandma and Grandpa Keller still lived in Fairfield. During Dad's surgery, Grandma stayed home at the parsonage in Fairfield giving a piano lesson, while Grandpa Keller drove over to Roland Hencke's to see how he was doing after his gall bladder surgery. Burnice, my dad's sister, waxed her kitchen floor, and then she darned Uncle Al's socks. Nobody was at the hospital except my parents. None of my mother's friends came, nor her sisters, nor any other member of the family.

Illness and surgery were mysterious and private things, not so much treated with denial by my family as with a collective unwillingness to spotlight the seriousness of such events. People kept their concerns and their worries at a distance—not absent, but challenging the immediate bearers of the burden to carry it unsupported. This was better than parceling out support from the beginning. That could come later, if it had to be that way. But how much better it would be, for now, if Mom and Dad could drive to Spokane, have the surgery, spend the necessary days in the hospital, and come home, on their own, to the casseroles the neighbors had piled high in the refrigerator. Anything more than casseroles—mowing the lawn, feeding the cattle, watching the kids—were promises stashed away like a secret bank account that one never wants to dip into, but will, if necessary.

The surgeon knew immediately that Dad was in trouble. People didn't tend to survive cancer in the fifties. With this in mind, the doctor cut out my father's thyroid gland, parts of his jugular vein, and a slew of muscles around the whole area. Now, this is what we call radical surgery, a shattering assault on the inner workings of my father's circulatory and immune systems, biorhythms, and general way of life. To say that neck of his hasn't been the same since is a profound understatement.

Mom had promised to tell him the results of the surgery as soon as she knew, and she did, leaning over his bed in that darkened hospital room. At twenty-five years of age, she delivered the news to him that the doctor had cut out his thyroid gland, and that, right then as she whispered to him in whatever postsurgery state he existed, his jugular vein was inventing new routes for his blood to arrive to his brain. Dad drifted back to sleep when she finished speaking, and Mom pushed her way through the air of her new life to the hospital cafeteria, sitting there by herself at that table for just a few minutes. Then she took her nickels, went to the pay phone, and said the word *cancer* aloud to the family members. They had all stopped their activities right before they imagined the doctor would finish the surgery and tethered themselves mentally to their telephones.

Our mother's obstinacy has driven us, sometimes, as adults, crazy. She categorically refuses to accept things she doesn't agree with and, in her motherly way, is quite capable of crippling each of us with guilt for daring to confront her perception of the way life should be. I believe, however, that this very same obstinacy has kept her, at certain times in her life, from cracking up. It was this obstinacy that propelled her to order the cup of coffee in the hospital cafeteria that day, that enabled her to put those nickels in the phone and repeat, over and over, the word *cancer* to family members, and that helped her drive back and forth to Deaconess Hospital day after day, with her young children screaming in the back seat as she navigated the icy winter roads. We now live in a world where getting in touch with our feelings is championed, but locking it all up inside kept her driving.

This approach to major problems has proven effective in the course of my mother's life and, luckily, she passed on the techniques to her daughters. I couldn't do the work of my adult life without learning well my mother's lessons.

* * * * *

For the past fourteen years I have been a teacher for children undergoing bone marrow transplants at a cancer research center in Seattle. People come here from all over the world for a chance to cure diseases that have resisted a long litany of failed protocols. Many of these people are children—in fact, schoolchildren. The treatment is long, four to six months, and the children receive schooling for the duration in whatever form their health allows. With a decimated immune system, due to the massive amounts of chemotherapy and radiation they have received, each of my students is a class of one, counting on me to drag them through the rudimentaries of world history, Spanish, DiNealean handwriting techniques, and multiplying fractions. Our ultimate goal is that their hometown schools will accept the learning they've done with me by granting school credit.

If my students don't die first, we are usually successful. I will not take "no" for an answer from a recalcitrant high-school counselor who can't see his way to granting a health credit to a teenager recovering from a bone marrow transplant. The credit is the challenge, not the rigid course requirements. You don't go through what these kids are going through and arrive at the other side without having learned more in this crash course than all the biology classes in China. I am only so-so at world history and DiNealean handwriting, but I am tenacious as hell when it comes to advocating for my students.

My students don't always survive, though, and when they die, it usually doesn't happen suddenly. My policy is to never cancel a school appointment just because someone is dying. Education really isn't so much about credit as it is about learning. The parents can cancel school as graciously or ungraciously as they can manage at this terrible time, but I cannot decide that a comatose kid can't still enjoy the exploits of Holden Caulfield in *Catcher in the Rye* or R. P. McMurphy in *One Flew Over the Cuckoo's Nest,* read aloud by me sitting in a chair next to the hospital bed, my voice a little bit louder than usual, competing with the wheezes and beeps of the machines that keep these children momentarily alive.

Reading aloud to comatose kids is not one of my favorite things to do. Nor is holding throw-up buckets for them while they vigorously toss the minuscule amounts in their stomachs for the fiftieth time that day or that hour. Nor is helping them operate the little suction machines that vacuum out their mouths, which are full of saliva and other bloody gunk that piles up because the sores lining their throats make it too painful to swallow. But I can do it because, among other things, my mom taught me denial. When reading, I deny to myself that this child whom I've gotten to know quite well during these intense weeks, is certainly dying. When holding the throw-up bucket, I deny to myself that his condition is any more serious than a bad case of seasickness. When handing a child the little mouth vacuum and watching her collection of bloody, infected mouth tissue shoot up the transparent hose to the jar by her bed, I say appropriate things (if I know the child well enough) like, "Oh, gross," as if she had just shown me an especially bad bruise on her arm.

One must develop a perspective on these things in order to make them acceptable. If I had attended each health event at my job with the degree of honor, seriousness, and absolute unacceptability that the event commands, I wouldn't have lasted one month.

* * * * *

For eleven days my dad lay in that hospital bed. My mother drove to the hospital, dropping my sisters off at Grandma's and picking them up again at night. The three of them returned home, the dot of the porch light a beacon to our house from the hill by Detlef and Mary's.

Mom arrived home to feed the cows, water the chickens, feed the girls, and shovel coal into the furnace. One morning she woke to completely frozen pipes. With the blowtorch and its instructions in hand, she struck the match, hoping she wouldn't blow up herself, her daughters, and the rest of the farm in the process.

It wasn't that there were no neighbors available to help. They offered, but she was of the mind to save those offers for the really hard times she couldn't quite describe to herself but thought she had to anticipate. Sometimes when you are in the middle of the hardest times you can possibly have, maybe the only thing that can help you is to imagine that they could be harder.

That winter was about defrosting frozen things, thawing out, hoping for the best, and imagining it could be worse.

It was almost Christmas 1952. Dad came home on Christmas Eve that year also. "Just like Jesus," I said once. Mom did not like that, thinking it not good to make sacrilegious jokes about something for which we should be so thankful.

For months after his surgery Dad had these huge, recurring headaches that he described as Fourth of July rockets blasting off in his head. It was a mystery that the Spokane doctors couldn't solve, so he and my mother took the train to Seattle, springing for a sleeper so they could pretend they were on vacation. The doctors ran a battery of tests and determined that the Lord only knew what was causing those headaches. Months went by, the headaches faded as mysteriously as the odd foot aches, and Mom and Dad chose to forget that Dad's top-to-bottom aches had ever occurred.

It wasn't until fifteen years later, when the rockets blasted off again, that Dad had one of those déjà vu moments, a moment right in the center of all that pain that told him he had been there before.

The bottom dropped out of the sugar-beet industry around 1910, propelling the area farmers who had put their stock and trade in this vegetable to switch to wheat or, as in the case of the Wakelys and the Uptons, two homestead families whose tracts butted up together and touched the Thams place on one side, to sell off parts of their land. They had to, in order to help repair the huge losses they had suffered. My Grandpa Hein, by then a man of thirty, bought a half-quarter from the Wakelys and an adjoining half-quarter from the Uptons, and asked my grandmother to marry him. He built our house as a wedding present for her, constructing it to rest on the line between the Wakely part and the Upton part, so that once the house was built, these two pieces of land united into one that was, of course, called the Hein Place. Our house is large and wooden, with two stories, a full basement, and windows all around. The gabled roof has a shed dormer on the east side and another on the west. The front door is off the large porch on the west end, even though almost everybody uses the back door by the privet hedge that stretches the length of the walk. Many a lovely sunset has been witnessed from that porch, accompanied by the pungent, sweet smell of alfalfa, especially in the spring.

Grandpa Hein was a sturdy man with a square jaw, a stern look, and a dry wit. He was lean and mostly known for his ability to fix anything. What was he like? As a child I was always afraid of him, although I have no idea why, except that his cigars smelled so bad and there was always one around, smoldering wet and gummy in that ashtray with the See, Hear, and Speak No Evil monkeys on the rim. He bought me my first Hula Hoop, his idea, and he gave me quarters on Flag Day, so he couldn't have been all bad. He also didn't seem to mind that my grandma always gave money to her younger brothers after they grew up and turned into no-goods.

In that Hein family photo, the oldest one we have, taken before Mary was married and when my Grandpa Philip was still in his late teens, he sits erect and stern next to Gottfriet, who is gnarled and bent, tucked between his eldest son and William, his youngest. Philip had a headful of straight brown hair, neatly parted in the middle. His bow tie is lopsided, and he looks awkward in what most certainly were his best, and most infrequently used, clothes. His hands sit uncomfortably in his lap, each hand seeming to scratch at the inside of the palm in a self-conscious way. There is no reason to think Philip is enjoying himself during this photo session, his face completely lacking the quizzical look of his father next to him or the amused stare of John, the fourth of the five children, who looks at the camera with an unexpected smile.

Gottfriet and Christina were, at best, in their early forties at the time of this photo, although each looked sixty, their youth betrayed only by their dark, grayless hair. Gottfriet's face is lined and dried, and his hands, resting in his palm, look permanently dirty and twisted, the way my father's hands looked when he was farming. Christina is a large woman, who stares stiffly at the camera, her hair in a tight bun, her mouth drawn in a straight line, her eyes squinting as if in front of some bright light. No one in the picture appears to want to be there.

Gottfriet and Christina sit there surrounded by their five children—the lovely Mary, soon to go off to marry Detlef; my grandfather, Philip, who married Marie Thams from down the road; George, who married Dolly Lundstrum and stayed forever on the Hein piece, passing it on to his own son, Clarence Hein, who now has passed it on to his son, Wesley Hein, my second cousin, who continues to live and farm on the original homestead piece. Gottfriet and Christina's other two sons, John and William, moved away and made new lives in other places.

Maybe I was afraid of my grandfather because of how he taught Dad to swim when he was a young boy: by tossing him into Fat's Pond. Grandpa himself couldn't swim, and he had a very hard time fishing out a sputtering, half-drowned Dad, who, terrorized, never was moved to become a good swimmer after that. Whenever I complained about having to go all the way to Spokane's YWCA for my lessons, Mom was quick to point out that they could always just toss me into Fat's Pond and see what happened.

Grandpa planted the willow, pine, and locust trees that now tower over our house. He drew "1917" and "Philip Hein" into the wet cement of the basement floor, right by the stairs that lead up to the outside. Those soft indentations just seem like irregularities in the concrete after all the years of being walked on by my grandparents and their descendants. But we all know what those dents mean.

If anybody was farm dirty, he or she came in through the basement. If they didn't want to take a shower in the concrete stall by the washroom, they at least had to wash up in the giant cement sink. My sisters and I often came in through the basement after playing outside. There was a grown-ups' closet and our kids' closet, each one filled with the appropriately sized winter coats, galoshes, and worn-out cowboy boots, all caked with mud.

Our farm lies between Spokane and Pullman, though it's several miles from the state highway that connects the two cities.

We are not in what you would call the heart of the Palouse but, rather, on the northern tip, much closer to Spokane than Pullman. Though Spangle is about eight miles away and Waverly only four, there is no doubt, even at a full ten miles of distance, that Fairfield is our hometown. Our address is a Fairfield rural route, our phone number is listed in the Fairfield section of the phone book, we went to school in Fairfield, and we were all born in the doctors' clinic in Fairfield. My sisters and I grew up with an allegiance to Fairfield, knowing in our hearts that it was the best place in the world to live, even if, as children, we'd never lived anywhere else and technically didn't even live there.

With its population of 500, Fairfield is now the largest community in the immediate area. The nursing home, up on Morrison Hill, marks the southern border. It looks out over the town, which sits mostly in a pocket among three converging slopes. On the eastern slope is the grade school, a red-brick two-story structure that has always seemed huge to me, even since I've lived in a city. Next to the school is the water tower. It has forever had the word "Fairfield" emblazoned in giant letters across it, lest any aviators have a doubt about their whereabouts. The John Deere dealership wraps around the northern town border. The front parking lot, full of big green machines, extends just a bit past the "Welcome to Fairfield" sign along State Highway 27, the main thoroughfare from Spokane to places south, such as Tekoa and Latah. On the western border is Zion Lutheran Church. The church's steeple competes with the water tower and the grain elevators in height. It adds a sense of solemn majesty to the town, a majesty that might be lost on a visitor, but certainly not on any resident. The railroad tracks cut through town, as well they should: they are the lifeline to the grain elevators. Fairfield was built essentially to complement those grain elevators. These giant concrete tube silos that tower above the city are full of wheat most of the year, or peas, lentils, and barley the rest of it.

Wheat means everything to us. Barley is important, lentils and peas have their place, everybody makes a little income from bluegrass and livestock—but our real meat and potatoes, in a manner of speaking, is wheat.

As far as we are concerned, civilization as we know it today, including everything from the ancient cities of Mesopotamia to the landing on the moon, we can trace back to wheat. If it wasn't for cultivating wheat, which freed people to do things other than scrounge for food, nobody ever would have had time to build anything short of a mud hut and gather that night's dinner. It was wild wheat, tens of thousands of years ago, that was the first plant to lend itself to cultivation. One thing led to another, and before long, farmers were actually able to grow and harvest more than they could eat, freeing people to build pyramids and acropolises.

Who would have thought, back then, that those scrawny stalks of wheat, poking up between the wild oats and thistles, would flourish into the world's most important crop, covering more of the Earth's surface than any other and producing about 600 million tons of grain each year, an amount that could fill a freight train stretching around the world about two and a half times?

Before Cyrus McCormick came along, a sickle was used to harvest wheat. Then Cyrus invented the reaper, which cut the wheat and tied it. Then the Pitt brothers, Hiram and John, invented the threshing machine, which shook the grain off the stalk. Then, in the 1830s, Hiram Moore and John Haskall came up with the idea of a machine that could do the work of both the reaper and the thresher. It could not only cut but also shake off the grain. This machine that so practically combined the two tasks of harvest was aptly named the "combine," and the world of grain production changed forever.

It wasn't until my Grandfather Philip was well into his life as a farmer that our family started using a combine. In Grandma's cardboard box of photos there are many images of the threshing parties they had before the combine. These large black-and-white pictures show teams of horses hooked to giant carts loaded with burlap bags, presumably filled with wheat. There are lots of people standing around—men, women, and children, many holding pitchforks. The thresher towers over them, as does the mountain of chaff it has spit out onto the field. Each farmer rented the thresher, which also included its operator, and all the neighbors came to help, as a group, moving from one field to another until the entire neighborhood's wheat was cut, threshed, bagged, sewn, and hauled away to the community silos by the railroad station.

Then the combine was invented, but still people shared these machines, because few could actually afford one. When my father was a teenager, Grandpa bought his first combine, which was worth more than the entire farm. With the combine, a single family could farm a thousand acres of wheat without much help, although my family never farmed more than six hundred.

Wheat is our thing, and a thousand acres of it swaying in the breeze is, for us in the Palouse, about the most beautiful thing on Earth. We put pictures of wheat on our Grain Growers Association calendars and write poems about it when we go off to college. We try to describe the soft lime green in the spring and its golden color in the summer, the way it rolls out over the hills in a carpet of agricultural wonder, moving slightly in sync with the breeze, and ripening in accordance with the sun. This occurs year after year after year in an awesome regularity that Palouse dwellers carry within them, even if they have moved away. I live in Seattle now, for reasons that sometimes I can't remember, and whenever I think of wheat fields, I feel a little soft inside.

The Palouse, an ill-defined area of farming country, is located mostly in Washington. Its southern border is unofficially the Snake River; the eastern border is the Tekoa Mountains in Idaho; the northern border is just south of Spokane; and a jagged line following the Channeled Scablands down between Sprague and Ritzville, a line including Washtucna, is the western border. If you don't know what a scabland looks like, just drive the twenty-five miles between Ritzville and Sprague and watch the country change from rolling green to dried ridges of craggy dirt. Then you will know.

I used to think *Palouse* was an Indian word meaning something spiritual, such as "Land of the Whispering Gods" or "Heavenly Skies" or something equally inspiring. And maybe it is so; or perhaps its derivation is from Palus, the name of the Indian group first mentioned in print by Lewis and Clark in their journals of October 1805. Or, maybe not so, since the other story of the area's naming was that French fur traders chose the name Pelouse, because it was the French word for what they saw in front of them: grasslands.

In Fairfield, there is no place where you can stand, by any one of the four "Welcome" signs, or somewhere between them, and not see wheat fields. Our town isn't just the buildings and the paved roads that run between them. It includes all the wheat fields that stretch between us and the other towns, big expanses of land plowed up a hundred years ago by our great-grandfathers. When a summer breeze comes up, it is hard to begrudge them, the grandfathers, their invasion of the landscape. The wheat, in all its golden wonder, leans and falls with each gust, and you know completely why anthems have phrases like "waves of grain" in the same verse with words like "majesty." No one would call Fairfield a majestic town exactly, but the very essence of the community is connected with this plant that surrounds us in all its majesty.

While Fairfield is just one of the many towns that dot the Palouse—towns with four or five streets, a handful of businesses, and a collective preoccupation with grain prices—it has the distinction of being the hub in our area of dot towns and fields.

It represents everything a town should be. It is small enough to see the fields and small enough so everyone knows everyone else. Most adults are members of one service organization or another, every child attends the same school, and anyone can accomplish almost anything he or she might need to do there.

Spokane is the nearest city, but we went there only to visit our grandparents, take swimming lessons, buy school clothes, and get library books. Everything else we could do easily in Fairfield. Mom would park the car on Main Street, giving each of us errands to run. This was her way of imbuing us with a sense of fiscal and familial responsibility. On any day, Marsha might head for Daniel's Hardware to buy flashlight batteries and WD-40, Cheryl to the bank with a check from Mom to get cash, and I to the drugstore to pick up Dad's refill and buy a pack of votive candles. Tracy tagged along with Mom, stopping at the butcher to get steaks from our locker, picking up groceries from Dodge's Thrift Store, and dropping off a receipt at the Pea Growers Association. We never needed any real money because all the stores kept accounts running—although once I got to charging too many cherry Cokes at the drugstore soda fountain, my dad did make me start paying for them out of my allowance.

Most of the stores are lined up on Main Street, one right after another, like a movie set or a toy town. On the north side of Main, the bank has a new building, a one-story structure with aluminum siding, beauty bark, and wheelchair access. They converted the old bank building with its plaster and rooftop curlicues into a restaurant that serves chicken-fried steak and reconstituted fish patties. The drugstore, another architectural feat of plaster curls, is connected to the old bank building. New owners have recycled old buildings into new businesses, with minimal cosmetic changes. The old blacksmith's building, with its anvil and horseshoe jutting out into the sky from the top of its square roof, has for decades been the newspaper office. The grocery store has always been on the same corner, with the same number of aisles and the same kinds of food in the same parts of the store. There is a small section for dry goods and a freezer locker in the back. It appears as if the same person who wrote the first giant shelf-paper signs in the windows is still writing them, the handwriting remarkably similar in today's weekly announcements to that in the signs in ancient grocery-store photos.

As long as I can remember, there's been a telephone perched on the windowsill by the front door of the grocery store, there for anybody to use, free of charge. I must have used that phone a million times over the years, to call Patty up the street to see if she wanted to meet me at the park, to call Dad at home for a prescription number, or to call Mom, once I got my driver's license, for a grocery list. I remember when Renee Clausen came running into the store one day, when Cheryl and I were in line for checkout, and screamed into the phone to her friend Janelle that Rick had proposed marriage. It was right out of the movies—Renee's tears of joy, everybody in the checkout line clapping, and Earl coming out from the back in his bloody butcher's apron to hand her two congratulatory steaks, all raw and bloody inside the newly wrapped white butcher paper. I remember wondering how she could be so surprised. Everybody knew she had been bothering Rick for two years for that proposal, and, as a young man with little spine, it was unlikely he wasn't going to cave in to her demands eventually.

Undoubtedly the saddest occasion that phone was used for was when Claudia called Dona that day Jimmy died out in front of the doctors' office years ago. His son, Greg, had rushed him into town from the field, hoping, of course, that the doctors would patch Jimmy together again, but he was dead by the time the two reached Fairfield. When Claudia called Dona, she didn't tell her Jimmy had died but, rather, asked if she could come out and tell Dona something. Dona had it all figured out before

Claudia arrived, since earlier she'd watched from her kitchen window as their pickup truck careened out of the field below at a speed not good for any part of the suspension system. Jimmy was always so careful with the machinery that it had definitely made her wonder.

The last time I was in Fairfield I noticed that Rick Kelly, who owns the grocery store now, finally traded the dial phone for one with push buttons. Rick inherited the store from his dad, Earl, who had come when I was a kid to work as butcher for Warren Dodge, bringing with him his wife, Goldie, and his two children. I believe Warren inherited the store from his father, who had bought it from William Cahill, the original owner of the business, who started the store before the turn of the century. The gravel road that leads up to our house was named after William Cahill, although I don't know why, since he never lived anywhere near our part of Hangman Creek.

If you cut the state of Washington roughly in half, leaving Wenatchee and Yakima on the western side, our eastern half is what you might call crowded with countryside. There are really only three urban areas: Spokane, thirty-five miles north of our farm, with a population of about 190,000; Walla Walla, ninety miles south, with 30,000 souls (not including the prison inmates), and the Tri-Cities area, a group of three towns about a hundred miles south, dedicated to the Hanford Nuclear Reservation. Altogether the three towns—Pasco, Kennewick, and Richland—have about 105,000 people. They don't make plutonium down there anymore, but instead spend our taxpayers' dollars cleaning up the nuclear mess they made during the forties and fifties.

The rest of eastern Washington is fields, plains, deserts, scablands, canyons, plateaus, and foothills—an area that in some ways has changed remarkably little since our great-grandparents came a hundred years ago.

First, there were the Indians. Then the pioneers came (my great-grandparents). Then their children married each other (my grandparents). Then they had children (my parents), who had us. And that is basically one way to tell the story of my neighborhood, where my father still sleeps in the bedroom he was born in on land that, in the history of time, has changed hands only twice: stolen once (from the Indians) and sold once (to my grandfather). This is countryside where many men still sleep in the bedrooms they were born in, on acreage that has never been sold. Most of our last names can be traced back a hundred years in the Palouse, to when our impoverished, desperate ancestors took a chance on making things better.

This is the natural way many places have developed all over the United States. On the opposite side of the spectrum would be Hanford, the most unnatural place in our neighborhood, although we don't like to think of it as in our neighborhood— and it really isn't, being more than a hundred miles south.

In the 1940s the Manhattan Project needed a place to make plutonium for atomic bombs. Among the handful of locations in which the government chose to carry on its war work was a chunk of land in the southeastern part of our state. This area, about fifty miles from the southwest corner of our Palouse, is icy in the winter and broiling hot in the summer—a dry, pockmarked landscape, ugly and craggy, with peeling layers of psoriasis-like dirt. The air swirls with gray dust and the fences are often packed with thistlelike tumbleweeds, slammed into piles and trapped on barbed-wire fences by the gale winds that frequent the area. Curling down from its starting point in Canada, the Columbia River makes its last U-turn right here before stretching west and flowing out a final 150 miles to the Pacific Ocean. It was a flow of water like the mighty Columbia that was needed to cool the plutonium.

Books will tell you that the Manhattan Project chose this area both for its proximity to the Columbia and for how remote and unpopulated the area was. In fact, the town of Pasco was quite prosperous then, by farming community standards, with an upsurge of population shortly after the turn of the century. In 1900 the town officially had 254 people; by 1910 the population had grown to 2,083; and by 1940 it was 3,913. Kennewick, across the river from Pasco and a scant ten miles away, grew from 183 people in 1900 to 1,219 in 1940. Richland, Hanford, Vernita, and White Bluffs were four little farming towns with main streets, schoolhouses, and grange halls, each with a few hundred residents. In other words, the area was a lot like our Palouse with its little towns and neighborhoods, although this neighborhood was clustered around the Columbia River instead of in God's best wheat country.

Farming was harder there. The soil is not as good as in the Palouse, and there is much less rainfall. Dry farming takes a different kind of skill, especially when your dirt is constantly blowing over to your neighbor's place. This part of Washington is known for its wind. The Columbia and its irrigating waters helped develop orchards and vineyards. Over the decades farmers had carved out homes, farms, fields, and orchards that, by the early 1940s, were beginning to pay off. And then the families got the letters.

On March 4, 1943, registered letters were delivered to 1,200 landowners in the Hanford, White Bluffs, and Richland areas, informing them that their property had been condemned as of February 23 and the families had sixty days to move. All they could know was that it was for the war effort.

It is a myth that, overwhelmed with a sense of patriotic duty and concern, the farmer families gladly handed over their land for the price the government was offering to pay. In fact, the farming families had a much saner reaction. They were shocked and furious. Emergency meetings were held, protests were organized, and, finally, the residents resolved to resist the money offered for their land, amounts that represented the naiveté or heartlessness of the eastern organizers of the sales. Ultimately, though, the families had no choice. Two thousand people moved in sixty days from land they had farmed for decades. It was months later that the monetary compensation arrived, these unsatisfactory prices invented by people who seemed to know little about farming and history.

The government letter these farmers received in the mail was officially titled a "Declaration of Taking" and precipitated the following collection of news items from the *Richland Courier-Reporter* in 1943:

A CALENDAR FOR DYING TOWNS

March 4: The letters arrive.

March 18: The Morey family returned to Richland to begin work on their farm. They had not heard of the evacuation order. They didn't even unpack their household goods, but returned to the coast where Morey had been working in the shipyards during the winter. The Pete Dennis family, also, not knowing of the order until they arrived in Richland, turned around and headed back to Iowa.

March 25: The Birthday Club of Richland had its farewell party at the home of Mrs. A. H. Willsen. Attending were Mrs. Bert Gray, Mrs. George Gress, Mrs. C. S. Billington, Mrs. A. S. Murry, and Mrs. E. C. Carlson. The club had been formed thirty-one years before.

April 1: After thirty-two years in business, George Gress closed the doors of his butcher shop, completely sold out of fresh and cured meats and other merchandise he had kept in stock, thus avoiding a prospective headache connected with the meat-rationing program that was to begin immediately.

April 1: The White Bluffs Grange held its last meeting, at which time the Charter was surrendered. About 150 grangers and their families held a farewell turkey dinner.

April 2: The White Bluffs Women's Club held its final meeting after thirty years of organization. About forty members attended.

April 15: The Hanford–White Bluffs commencement exercises originally scheduled for May were held in the Hanford Auditorium. Seven seniors from White Bluffs graduated, two from Vernita, and six from Hanford.

April 18: The board of the White Bluffs Cemetery Association opened its meeting to all who had friends or relatives interred in the cemetery. Edmund Anderson was authorized to find a place in some nearby cemetery where graves would have perpetual care.

April 19: The Richland Women's Club met for the last time at Vic Nelson's home. Moneys in the treasury and from the sale of the club rooms were donated to the Red Cross, War Relief, Salvation Army, two children's homes, Children's Orthopedic Hospital in Seattle, and war-orphans organizations.

April 22: The Hanford Grange regalia, flag, and song-books were donated to the Benton County Pomona Grange. The Hanford Grange had been organized in 1919 and had nearly a hundred members. It owned twenty-one lots, including the Grange park on the shore of the Columbia River, and the Grange buildings.

April 23: Richland schools held their last classes. Seniors were handed their diplomas without commencement, baccalaureate, or any customary exercises. There were twelve graduates.

April 29: Richland's Social Hour Club held its last meeting. It had been formed early in 1934. The club promoted sociability, neighborliness, and kindness, and in so doing, financially aided Pasco Hospital, the Children's Home, Children's Orthopedic Hospital, the Richland park, and many other worthy causes. It donated its cash on hand to the Red Cross.

May 3: Arnold Willsen died at his home in Yakima. He had operated a blacksmith and machine shop in Richland and had lived there for many years. He had moved to Yakima just a week before his death.

May 6: One hundred seventy-seven caskets from the White Bluffs Cemetery were moved to a section in the new cemetery in Prosser. Families bought the plots, but the government paid the moving expenses.

May 15: Parents were urged not to mention the displacement to their sons in the service so as not to demoralize the troops.

May 20: The last meeting of the Up-River Women's Club was held at the Claude Jones home in White Bluffs. Accumulated funds were sent to McCaw General Hospital in Walla Walla and Children's Orthopedic Hospital in Seattle.

May 20: The Hanford–White Bluffs combined high school and the grade schools at Hanford, White Bluffs, and Vernita were closed.

May 27: White Bluffs, supposedly named by the early Indians after the white clay banks along the east shore of the Columbia River, will soon be a forgotten town. The entire community, consisting of 193,833 acres, must be vacated by May 31. Many families have made White Bluffs their home since 1892.

June 3: The *Benton County Advocate* ceased publication. Paid-in-advance customers received the *Kennewick Courier-Reporter* to finish out their subscriptions.

By the middle of June, the towns of White Bluffs and Hanford no longer existed. The town of Richland was disincorporated and placed under the administration of the Hanford Engineer Works. The school districts of Vernita, White Bluffs, Hanford, and Richland were consolidated as the Richland School District No. 400, and all records were transferred to Richland. Prisoners from the penitentiary in Walla Walla were brought in to harvest the orchards and vineyards that were taken away from the farmers, a harvest that continued for several years after the war.

Where the town of Hanford had been, a giant camp was constructed to accommodate the 50,000 people who streamed into the area to build. They came from all over the country, responding to leaflets and newspaper ads in large- and small-town papers. The camp became a small city, complete with barracks, mess halls, and other facilities, including a women's section that was fenced off and guarded at all times. The secrecy, isolation, wind, and dust were demoralizing to the workers, and the government contractors offered recreation programs and good food to help keep their spirits up. Twenty-five million meals were prepared during the tenure of the Hanford camp, fifty tons of food for each meal.

And then, after just a year and a half of intense building, the construction was completed. In this short time, workers had built more than 500 buildings, 158 miles of railroad, and 385 miles of road, and poured 780,000 cubic yards of concrete. In the entire time, nobody, except for a few, knew what they were working on. A sort of convoluted patriotism prevented much speculation. A few people were needed to keep the secret, and many were needed to work in the dark.

Richland had been rebuilt to house the masterminds who would work at the plant. The town was entirely government owned and controlled. "Alphabet houses" were constructed, clustered into neighborhoods according to floor plans. The ABCs were on one hill, the EFs on another, the HKs across the way, and the QRSs in a different part of the new town. Ten different floor plans were used, with identical postage-stamp lawns planted in front and two sycamore trees per front yard, irrigated with water from the nearby Columbia River. The trees are tall now, and people I know who were children there say things like, "Oh, yeah, I grew up in a Q."

Richland was not a normal town, especially for rural eastern Washington. Almost every resident was a government employee in one way or another. Everybody worked in secret jobs, lived in look-alike houses, and were strangers from somewhere else. Nobody knew or cared much about the land that surrounded them. By 1950 the population of Richland had grown from a few hundred to more than 15,000. Pasco and Kennewick, where many of the workers lived, had grown respectively to 10,228 and 10,106.

In 1945 the first atomic bomb was tested in Alamogordo, New Mexico, and one month later President Truman ordered the bombs dropped on Hiroshima and Nagasaki. In each city, more than 70,000 people were killed in an instant, and the war was over.

The *Pasco Herald* put out an extra edition on August 6, 1945. "It's Atomic Bombs!! President Truman Releases Secret of Hanford Product," the headline screamed. The front page was littered with quotations from residents. "If you don't get in the Army, you can hope to win on the job here," said Kenneth Jensen, who worked on the project for two years. "Finally," said Mrs. S. F. Schiecht of Kennewick, as she waited for the hardware store to open. "I've never seen a job where things were kept so secret. My husband works there, but he didn't know anything about it." Dell, the night clerk in the Transient Quarters, refused to give any more than his first name or discuss the matter at all, beyond stating, "I've been secretive for two years. Why should I change now?"

It was a headline in the *The Spokesman-Review* that informed my family that both the bomb at Alamogordo and the bomb that was dropped on Nagasaki contained plutonium produced at Hanford. That's how everybody—everybody in the whole world and everybody in our neighborhood—found out what was going on down there: from the newspaper, after the fact. People in all of eastern Washington took a certain pride in the part they had played in ending the war, assuming that part of ending it was living beside the secret.

"I guess it was more people like me," Leonard told me years later, "soldiers who actually saw what we did to the Japanese people in Hiroshima and Nagasaki, who couldn't feel so certain we were heroes. And at the time we had no idea what we were doing to ourselves here at home. If we'd known that, I think we would have felt a whole lot different about this plutonium business."

My mother, Dolores Keller, arrived in the Palouse as a teenager from North Dakota, a preacher's daughter. Mom was the oldest. Like the rest of her family, she's short, a little over five feet tall, with a compact body that moves fast. All the Keller women talk fast, move fast, and can handle several projects at once.

Mom spent her childhood moving from one North Dakota country parish to another, living in towns of twenty or thirty people and sometimes fewer. As the oldest preacher's kid, she was held up constantly as a community example for how to raise a devout, educated, and dedicated daughter. In other words, my grandparents raised my mother to be an obedient prude. She was taught early in life that passion was a loss of control, that athleticism was unladylike, and gossip was un-Christian. Not only were the eyes of the Lord on her always, but also the eyes of the parishioners, who were quick to criticize the preacher if his own children strayed from the family hearth.

I don't think my mom had much fun as a kid. They moved too much, and whenever they stayed places, her peers probably resented the pristine upright figure she was destined to cut. It wasn't that my mom's people lacked a sense of humor, it's just that they put everything through the God-filter, which has a tendency to cut down on spontaneity. This can be especially stultifying in the growing-up years.

While my mother was away at college—Pacific Lutheran College, of course, in Tacoma—Grandpa Keller received the call to be pastor of the Lutheran Church in Fairfield. As a preacher's daughter, Mom had learned that it is better to be attached to ideas and not to landscapes, as the latter would inevitably change when a parish needed new sermons. Mom went off to college when her hometown was Colfax, a wheat-farming community in southeastern Washington. While she was away at college, home became Fairfield, another wheat-farming town sixty miles to the north.

When, after two years of education, Mom needed to postpone college to earn tuition money, Fairfield's bank president, a deacon in Grandpa's new parish, saved a job for her. It was a sign of good Christian faith to hire, sight unseen, the oldest daughter of the new Pastor Keller.

Mom met Dad when he came in to do his banking. He was home from the war and, feeling worldly, didn't mind taking on a college girl. My parents would never breathe a word to us about the length of their courtship, but we know it was short because of how adamant Mom is about not telling us. In this particular case, she considered herself a bad example for her daughters. Even to this day, as we are all well into our adulthood, with a miscellany of marriages and boyfriends under our belts, does she refuse to tell us. It is her big secret, and I can easily imagine one of us desperately trying to extract it from her on her deathbed. "Please! Mom! Wait! Just tell us how long you knew Dad before you married him!"

Serving as a bad example was as bad a sin as any conceived in the black-and-white world of my mother. Things are either right, or they are wrong. Setting a good example is right, going to church is right, giving money to the poor is right, and doing volunteer work is right. Their opposites are wrong, some more than others, but all falling into the category of completely unacceptable. It's no wonder that in our adult lives all of her daughters have routinely moved back and forth between the right and wrong worlds of our mother, trying to find some happy in-between place that accounts graciously for human foibles.

Some people might describe Mom as strongly principled, one you can always count on to take the moral high ground. Others might describe her as rigid and intolerant. I have described her as both, depending on whether she has sent me a check to balance equally the money she has just sent my sisters, or whether she has announced that no unmarrieds can sleep together in her house without the institution of marriage, no matter how old they are. Always she adamantly insists that Dad feels as strongly as she does about these ethical questions, although we doubt it. He supports her, we think, more out of a sense of strong loyalty to a wife who has stuck with him through thick and thin.

Mom was twenty and Dad twenty-seven when they got married. He was a handsome guy with dark, wavy hair combed back from his face, his hairline just barely hinting its future recession. At five foot eight, he wasn't tall, and at 170 pounds he wasn't fat. He worked hard and was athletic. The locals generally thought of him as kind, funny, and honest. He liked to drink now and again, dance now and again, and have fun playing sports and hunting. He did go to church, but it was mainly to take Grandma Hein, because Grandpa refused church attendance and Dad didn't want his mother walking in alone. It just didn't look right.

Mom always told us she married Dad because he took his mother to church and because, on their first date, he brought her roses. I like to think it was because he was handsome and could do a mean jitterbug, a quality my mother had learned to appreciate as college had helped her shed some of her prissiness.

Mom herself wasn't half bad looking. She had high cheek-bones and long, blond hair that rolled away from her face in the fashion of the day. She has always been high-strung, opinionated, attentive to the details of people's conversations, and full of laughter. She likes to sing, play the piano, and read. Her parents never once had allowed her to dance the jitterbug, but she spent two years in college on the coast, a startling accomplishment among the wheat-country girls of her generation.

Mom and Dad were an unlikely pair, but with the postwar euphoria of the times, they threw caution to the wind, married, and took over the farm. Grandpa and Grandma Hein moved to Spokane. Mom became a housewife and joined Homemakers, and Women of the Church. Dad joined Toastmasters and the Service Club, played town team baseball, and was on the Fairfield bowling team.

When Grandma and Grandpa Keller got their next call to a town in Oregon, Mom took over playing the church organ and teaching piano in a tiny room at the top of the grade school. She charged $2.50 an hour for lessons. Mom cooked huge meals during harvest and learned to drive a truck. On weekends Dad taught her how to dance some, although both agreed she never got good at it. Dad says he quit dancing because Mom didn't want him dancing with women who were better dancers, and Mom says Dad quit dancing because his leg quit working and, like everything, if he couldn't do it well, he didn't want to do it at all.

As my grandparents' only son, Dad never considered any profession other than farming. To this day there are parts of our land for which money has never changed hands. Abe Lincoln gave it to us and our neighbors free of charge, and parents have been passing that acreage on to farming children ever since.

Luckily, Dad was pretty good at farming. Not that he would ever say so, of course, but in 1966 he won the Inland Empire

Conservation Farmer of the Year Award for our district. The Inland Empire, which used to be called the "Great Columbia Plain," is the area framed by the Cascade Mountains to the west, the Rocky Mountains to the east, the Wallowas in northeastern Oregon, and the Canadian border.

The local newspaper, the *Standard Register,* sent photographers to our farm when I was twelve. The photos portray the Heins as a model farm family, showing us lined up in front of the barn and also down at the creek. Luckily, my cousin Mike, who sometimes spent summers with us, got to be in the pictures, which helped hide the fact that our model farm family didn't have an obvious male heir to the estate.

A good farmer is like a good hunter, a good doctor, or a good artist. Education is one part, whether it comes from an Ivy League university or someone's father, but the real mark of a good farmer is a certain finesse, a sense of timing, an ability to know when. It's a thing that some people have inside, and it's not as simple as it looks.

Imagine that you have a few hundred acres of land. It has a certain kind of soil and access to a certain amount of water. You must know what does, or doesn't, grow in this soil, in this climate from season to season, and which crops garner more money than other crops. The climate and the money change constantly. Crops must be rotated among the fields, peas one year and wheat the next, to prevent the same nutrients from getting sucked out of the soil year after year. Sometimes you must let the whole field lie fallow for an entire season to build up the dirt, like a field sabbatical. Some fields need to lie fallow more often than other fields. Also, you'll receive the occasional letter from the government, which, in the interest of keeping the market from being flooded, is now offering you money not to plant certain fields. Most small farmers farm not only their own land, but often that of other people who have nearby pieces. Keeping them happy while balancing government subsidies, grain prices, and seed bills adds another dimension to your project.

Once you've decided which crops to plant where, and what to leave fallow, you have to decide when to plant. The farmer bases this decision on the weather, factoring in that, before planting, you have to plow, spread manure, and disk.

Then you have to hope. You have to hope that there is enough rain. You have to hope that there isn't too much rain. You have to hope that it doesn't freeze too early. You have to hope that there isn't hail when the seedlings are still small. You have to hope for enough snow to cover the winter wheat, but not so much snow that spring brings a flood. One year we had to hope that the ash from the Mount St. Helens eruption didn't choke the plants.

Then you have to harvest, hoping that the weather cooperates. Because different crops ripen at slightly different times, but all in a span of approximately a few weeks, a farmer must weigh his determination to harvest things as they ripen with knowledge that another crop in another field is on its shirttails. It's a balancing act.

Harvest is the monumental event of the year. I remember huge, hard days that were long started when I arrived at the breakfast table and ended when night came again. God and the weather said, "Now's the time," and everybody shifted into action. The hired men lived in the basement, where they slept on cots and used the old concrete shower and an ancient toilet that leaked all the time. Russell had been my grandfather's hired man and continued helping my father. He lived with us also during harvest but slept upstairs on a cot in the hall. Neighbor boys were often our hired men. These boys came from families with one or two sons to spare, their parents passing them on to us when the younger brothers were of an age to begin helping their own fathers. It was a rite of passage. A farmboy works for his father first. When his younger brother comes along, the elder must learn to work for someone else. If the eldest son inherits his family farm, his teen years may be the only time in his life he works away from the family farm. It's his university of sorts, learning what he can about farming from his father and then moving on to a new mentor. It gives perspective. My father was a mentor to a number of neighbor boys.

From dawn to dark they harvested. Dad drove the combine round and round in monotonous oblongs, over and over, up and through the draws to circle the hilltops and hang off the sides. His combine, old and perennially patched together, didn't have a cab but, rather, an open seat. Dad sat on the metal seat with a kerchief around his mouth to ward off the flying dust, chaff, dried weeds, and general harvest-airborne rubble. The hired men would make trips back and forth in the trucks between whatever field they were harvesting and the Grain Growers Association, where they dumped the load of peas, lentils, wheat, or grass seed to be weighed, gauged, and recorded under the name of Ralph Hein. It was hot, dry, and filthy.

Some women drove the trucks, but mostly they cooked huge meals that sat warming on burners until ten at night. They prepared platters of fried chicken and deep bowls of steaming mashed potatoes, filled a gravy boat to the brim, heaped a bowl with mixed vegetables, and dished out a jar of applesauce. They always had a loaf of homemade bread on the table and some kind of pie with ice cream waiting. The wives made everything fresh that day, including the bread. A harvest crew is like a plague of locusts after a long day of work.

This harvesting and eating goes on for weeks. Finally, in late August, it ends. The combine is put in the shed for repairs, the hired men move out of the basement, and the wives take the extra leaves out of the supper tables. Then the farmer must again begin to think of orchestrating the symphony of the coming farm year while deciding when to sell this year's crops.

The going price for each crop fluctuates all year. Dad's crops of wheat, peas, lentils, and grass seed would sit in a giant silo at the Grain Growers, mixed in with every other neighborhood farmer's crop of the same pedigree. The farmers would wait to sell their bushels to Russia or places equally mysterious. They waited for a good price, or maybe a mortgage payment to come due, or just the last possible minute, always hoping the price on the world market would creep to a new high. When the spirit moved him, Dad would pick up the phone and say, "Sell." That's the final step in being a good farmer. You have to sell at the right time.

Winter was for fixing farm equipment. My father inherited my grandfather's ability to keep ancient trucks, combines, disks, tractors, swathers, and manure spreaders alive, not to mention the family car and his pickup. He also raised livestock on the farm. We had chickens, horses, and cattle, as well as a dog or two, and an uncertain number of cats. Not only could Dad tune up a combine and weld a broken swather, he could butcher a steer, deworm a horse, and behead a chicken without batting an eye. I grew up assuming everybody's dad knew how to castrate a calf.

My father's tremendous desire to produce a son before he traveled to the hereafter directly relates to my presence here. Can you imagine the pressure he felt? I have often said that if it wasn't for cancer I wouldn't be here. There he was, thirty-two years old, the only son of a farmer, tending the land of his parents, who had farmed the land of their parents, who had homesteaded a major part of our property before Washington was even a state. And here my father was with a farm, a wife, two daughters, and cancer. This was why, with the scar on my dad's neck just beginning to heal, and neither of my parents knowing whether Dad would be alive or dead in a year, they decided to have another baby.

I was born on my mother's twenty-sixth birthday. Dr. Hart and his nurse, Sunshine, sang the birthday song as Dad helped Mom into the clinic at three-thirty that morning. I have no doubt that in the moment I arrived there was considerable disappointment regarding my gender. Yet I have never, except perhaps a couple of times during high school, felt any serious regret on my parents' part about my existence.

I was born in 1953, a time of general world paranoia. Senator McCarthy was telling everyone that not only famous movie stars, writers, and artists were Communists, but even the next-door neighbor could be one. Communists were bent on the destruction of our democratic way of life and were pursuing this goal by building bombs big enough to blow up the universe and those of us in it.

Clearly, the United States had to build bigger and better bombs. One couldn't say we in Washington weren't doing our part, with Hanford just down the road. Blowing up Nagasaki, as it turned out, was just the beginning. In the fall of 1946, General Electric took over administration of the Hanford Reservation, and a new wave of construction workers migrated to the area. After the bomb was dropped and the war ended, the residents just assumed the Hanford Project would fold up, but now it began to appear that bomb making and other secret missions were the area's future.

In 1947 the expansion of Hanford began, including the construction of five new plutonium-production reactors, two chemical-reprocessing plants, and eighty-one underground waste-storage tanks. In 1949, when the Soviets detonated their first atomic bomb, no one could begrudge our nation's commitment to keep ahead of them in matters of defense.

The wish to ensure protection from our enemies was not without its risks. Mistakes were made, and radioactive wastes were periodically leaked into the atmosphere. The most predominant releases were of a radioactive waste product called iodine-131. Maybe you could look at it as nuclear exhaust allowed to escape out the back of a nuclear plant, presumably in the name of powering forward. Before they figured out how to filter it, the iodine-131 went into the air, traveled with the winds, and landed wherever the wind died down. If the wind stopped over a grass field, of which there are many in the agricultural areas of the Northwest, the unsuspecting milk cows that grazed that particular field, in a tribute to twentieth-century science fiction, produced radioactive milk, although no one realized this at the time. People who had their own family cow were, in some respects, at more risk than people who bought their milk. Purchased milk, even if somewhat tainted, was invariably mixed with a certain percentage of untainted milk, which served to dilute the toxicity level.

The Hanford scientists released the most significant amounts of iodine-131 between the years 1944 and 1947. After that, they figured out how to install filters that captured most of it, and, short of a failure or two in the filtering system, managed to contain most of the deadly material before the winds swept it across the area. Ironically, in the late 1940s, specialists urged mothers to replace breast-feeding their babies with cow's milk.

My sister Marsha was born in the late forties with an allergy to any dairy product that wasn't homogenized, so Dad sold our family cow, Betty the Guernsey, who for years had been producing the accompaniment to his Frosted Flakes. Mom took up shopping at the Milk Bottle in Spokane. This dairy sold milk in returnable half-gallon glass bottles with thick wire handles for easy carrying. The Milk Bottle, a two-story concrete building shaped like a giant milk bottle, is still there today, although I believe it is an insurance office now.

* * * * *

The fifties were complicated times for my family. On the one hand, my father had demonstrated a propensity for cancer, an illness people were unlikely to survive in the 1950s. Living on the family farm further complicated this issue because now the children of Ralph and Dolores were three girls, and girls did not often take up farming. On the other hand, the Russians could blow us all up at any moment, making concerns for keeping the farm in the family rather a moot issue.

My mother is both practical and optimistic. She was optimistic that we wouldn't be blown up anytime soon, and practical about finances. If Dad were going to die, she would need to make money, more money than she could earn as a bank teller or piano teacher. Collective paranoia about the bomb only fanned Mom's worries. She decided to give the boy thing one more try and then go back to college. Enter Tracy, my little sister.

All four of us were born in the Fairfield Doctors' Office, delivered in the middle of the night by Dr. Hart and Sunshine. They delivered all the town babies. This was a great convenience, because the city hospital, an hour's drive away, was prone to keeping women overnight, generally regarded as an unnecessary fuss by my parents.

The labor pains would start, and off Mom and Dad would go to Fairfield after a phone call to Dr. Hart's house. Mom gave birth while Dad sat in the waiting room. Then, all tuckered out, the newly enlarged family drove home. The grandmas descended, making big pots of knoëdel and shrimp-wiggle. The neighbors stopped by for a peek, and so would the hired men. Before heading to the fields they came into the house, trying to walk softly in giant work boots. Most of the hired men didn't actually touch the new baby, but stood back with arms crossed, nodding shyly. Of course, Russell, the hired man who was like a member of the family after two farming generations, would immediately pick up the new baby. He was like another grandfather and started his surrogate duties by scooping up each of us with determination, but gently, on that first day of our existence, swaying us back and forth, saying soft words in his Bull Durham breath. Meanwhile, Dad placed the spotlights around the living room, and the movie camera rolled. The grinding sound and the hot lights usually caused the new baby to whimper, if not to weep. Too bad for baby. It was all part of being born into our household.

Tracy was, in some ways, the biggest surprise of all. Mom and Dad just couldn't believe it. They'd chosen the name Bart Philip, fixed a room upstairs, and filled it with overalls and John Deere toy tractors. Tracy swears to this day that any emotional problems she may have are because we treated her arrival into the world with such great disappointment. This was not true. I was only six, but I remember the day as one of great elation and good jokes. Most people don't realize it, but German Lutherans can have quite a sense of humor, albeit subtle. We can make jokes about anything, including the birth of a fourth daughter to a farmer with a susceptibility to cancer.

Any emotional problems Tracy may have today are not due to her gender, but more probably because I tortured her mercilessly when she was younger. Oh, not at first when she was just a baby or even as a toddler. She was too cute for torture then—extremely cute, with her golden, happy-girl demeanor and her white-blond hair that curled all over her head like she'd just come in from a windstorm. She had striking blue eyes and an open, gregarious personality. I dragged her around by the hand at all public events, prompting her to say each new word she learned for my friends, picking up her toddler self and hauling her around with exaggerated familiarity, spitting on my fingers and wiping the corners of her mouth the way that Mom did mine when it was crusty.

I don't remember when exactly I turned on her, but I know she celebrated when I went off to college. Even now, when she and I spend too much time together, we revert to what Mom calls our old patterns. I pick up on her tension about something and then, in a cheap sinister way, make it that much worse for her by playing upon it. I can't help myself. Just last Christmas was my last Tracy Torture Fest.

For as long as I can remember, Mom has always made Christmas dinner for our large, extended family. This past December, right in the middle of opening presents on Christmas morning,

Mom became very ill, just hours before twenty-two people were coming over for dinner. This could have been considered catastrophic except that Mom did have, for once, all four of her adult daughters at home. We could pull off this event, especially since Mom, the quintessential planner, had most of the food prepared already. It was just waiting in Ziploc bags downstairs in the basement freezer, frozen into yellowish yam lumps and creamed-onion bricks.

With Mom out of the picture, it was astonishing how quickly her daughters stepped into their historical roles. Marsha became the greeter, welcoming people into the house, taking their coats, marveling at how this or that child had grown, and generally making people feel as if we were happy they were there, even though we sort of weren't any longer. Cheryl and Tracy charged into the kitchen, determined to raise dinner out of that pile of frozen, full Ziploc bags.

And what did I do? Well, first off, let it be known that we have a very small kitchen, at least the area where the stove, sink, refrigerator, and cutting boards are. There is a larger area with the breakfast table, but the part of the kitchen that has to do with preparation is tiny. It's hard for more than one person to maneuver in there, and nearly impossible for two. Weighing the reality of this situation, I decided I might as well stay put, out of the way, sipping a glass of wine and chatting with anyone who had tired of Marsha. I made this decision fully cognizant of the fact that I would be first in line when cleanup detail kicked into gear. While everyone else digested away, I would be scurrying around, scraping scraps, scrubbing counters, and jamming one more fork into the dishwasher before starting the first load. I loudly voiced this plan periodically, just so no one would think me a slacker. Usually Cheryl, who is really quite good-natured, would say, in between slamming the oven door or zapping a yam bag in the microwave to get it thawed, "Oh, sure, that sounds familiar," as if she thought I wouldn't really help, but she was just kidding me. She knows that I usually do, in fact, the majority of cleanup on Christmases.

Tracy, on the other hand, was not handling the demise of Mom gracefully, or at least in the same context as I. Tracy is a disaster relief coordinator by trade, which means that if there's a disaster in her California county, it is her job to organize the forces necessary to waylay any potential death and destruction. It's a fairly serious job, and I'm sure she does it well, if watching her switch into her disaster mode on this Christmas afternoon was any indication. Her mission, that day, was Christmas dinner. In order to reach this goal, she was completely focused on getting that food on the table, getting those people fed, getting dessert in them, and getting them gone. Job done. Another disaster avoided.

"Do you think you could find it in your heart to, maybe, *help*, Teri?" she suggested sarcastically, clearly having ignored my announcements about my future cleanup plans. Noting at this point that there were three people in the minuscule kitchen area, their contortions to get what they wanted a bit like playing a game of Twister, I said, "Just what would you like me to do, Sweet Thing?" I know Tracy hates to be called "Sweet Thing" by me. I've always called her that in a way that she thinks is insincere and mocking her. She is completely correct. "Well, you could set the table!" she snapped. I looked again into the kitchen, where the Twister game now more clearly resembled a telephone booth of human bodies. Apparently, learning of Mom's retreat to her sickroom, our guests had decided they needed to pitch in this year. There was no shortage of able-bodied helpers for this Christmas dinner. In fact, there was, by the looks of that kitchen area, if anything, a dangerous surplus. The plates and silverware were on the far side of that squirming mob. Since we weren't supposed to eat for another hour, it seemed practical to wait until the area had cleared out some before I tried to locate anything.

I poured myself another glass of wine. I would have explained this situation to Cheryl, had she been the one to ask me to set the table. That's because Cheryl, as usual, was cheerful and nondemanding. For some reason I just didn't think I owed Tracy an explanation as to why I wasn't jumping into the fray of that kitchen. Why play into her control issues? Why succumb to her railing, just because she couldn't relax about dinner? Wasn't this a day for all of us to enjoy, including me, who was happy to enjoy, and her, who was incapable of enjoying it? Besides, she went to college. She should be able to see why I wasn't getting the damn silverware. It was a madhouse in that three-by-five cubicle!

"Would you get off your ass and do something!" came a thunder from the middle of the living room. A silence fell over our guests. Tracy was losing it. I glanced around and noticed that her young children were still having their naps. Thank God they weren't exposed to their mother's blasphemous loss of control! She stood there in the middle of the living room, giving me the "screw you" vibes, although she had regained enough control to just vibe, rather than utter, something that would have been unheard of in our house. This was Christ's birthday, and our house was swarming with Lutherans! Even saying "ass" went way over the line. Thank God Mom was in a nonprescription-drug-induced sleep in a far corner of the house!

"Now, now, now," Marsha said, as she had said a thousand times before when we were growing up, "Christmas is no time for such bickering."

"Well, I never said anything," I countered. "I'm just sitting here drinking away, minding my own business, when I am brutally, verbally attacked by Tracy! I feel cheated out of a happy Christmas!"

"Number one, Teri," Cheryl decided to get involved, "you have never minded your own business in your life. I don't imagine you've decided to start today. And, number two, let's just get dinner on. Tracy, get a grip."

Right then I noticed that Tracy had tears in her eyes, and I realized a sad truth. I wasn't explaining my reluctance to get the silverware because I wanted to help her with her control issues but, rather, because of some sick pleasure I got from torturing her, knowing it would get to her even more if I treated her anxiety about dinner with passivity. Damn, now I felt sort of guilty and realized, with a hint of sadness, that my pleasure at torturing her wasn't nearly as satisfying as it used to be. In fact, as an adult, when Tracy wasn't incredibly uptight about something, a state that can happen to her a little more frequently than the rest of us, although not that much more, I actually enjoy her company. She's very good-hearted, has darling children, has a husband from El Salvador I can practice my Spanish on, and has a lot of the lightheartedness of Marsha. To her credit, though, Tracy would never sell Amway or do any of that cult stuff we're always afraid Marsha's about to spring on us. Maybe my days of torturing her are on the wane.

The worst thing I ever did to her, or the best, depending on how you want to look at it, occurred when Tracy was nine years old and I was fifteen. She and Mom had just returned from a trip to Fairfield. I heard her run upstairs, where I was in my room probably reading *Soul on Ice* or listening to Iron Butterfly. I noted the unusual way Tracy ran up the stairs, darted into her room, and immediately shut the door. I was suspicious, so tiptoed down the hall, took a deep breath, and swung open her door without knocking. She leapt away from her bed and started yelling, "What do you want? Get out of here!" as she wedged her body between me and the bed. I was bigger than she was, so she didn't have a chance. I pushed her aside and yanked back her bedspread. Underneath was a Fort Knox–like cache of candy.

"I hate you! Leave me alone! I bought it!" she stuttered, knowing she had been caught by the sister least likely to sympathize with her desperate need for candy, a desperation that until only very recently had been thwarted by her lack of money. There was enough candy there to stock a healthy Halloween bag.

In a way, I felt guilty for the pleasure I took in informing our parents of my baby sister's transgressions. Mom marched her youngest daughter and her bag of booty right down to the car and drove back to Fairfield, where Mom dragged Tracy into the drugstore and made her declare her crimes to Ray Goodner, the druggist. Ray was also the father of her classmate, Sam, whom she'd always had a kiddy crush on, and who would certainly hear all about this at the Goodner family dinner that night. In fact, I still feel sort of guilty about it, knowing how utterly miserable Tracy was for at least two days as she slumped around the house like a wretched felon. But also, when I think of it, to my knowledge she never shoplifted again, or at least got caught for it, which means I either saved her from a life of crime, or I helped her polish her technique.

I don't know why I was so hard on her. Mom says it's because Marsha and Cheryl mercilessly tortured me, and I had to pass it on to someone. I don't remember being mistreated by them, but I like to think this is the reason. It pleases me to consider that my faded memory might be a replica of Tracy's. I wish she would forget those years of torture and, particularly, my finking on her during her Fairfield Drugstore Candy Heist.

* * * * *

Shortly after Tracy's birth, Mom went back to college. I grew up with a mother who, as an English major, wrote papers on Dante's *Inferno*. She was forever trying to engage us in discussions about the writings of Thomas Wolfe and F. Scott Fitzgerald.

Pilgrim's Progress was fascinating to her. She alternately read John Milton and C. S. Lewis, with a bit of Robert Frost for variety. It must have been a bit frustrating for her. My father read *Sports Illustrated* and *Time*. When we girls read, it was usually Nancy Drew or Walter Farley. I don't know what the women at the Homemakers Club were reading, but it probably wasn't George Eliot. Mom kept at it, though, typing her essays on literary themes late into the night on the erasable bond paper she bought in reams.

That was the good side, I suppose, if you consider *Pilgrim's Progress* a good bedtime story. The bad part was that Mom often was not home, and when she was home she was doing her homework. This meant we were to keep the television down, we were not allowed to wail with Tarzan, and we had to try our damnedest to keep Tracy quiet.

Mom and I started school the same fall, she at Whitworth College and I at Fairfield Elementary School. I don't know who was more excited. I was about to have my first teacher, and she was about to learn how to become one.

We didn't have a kindergarten in Fairfield. I arrived at first grade with a vague idea of my letters and an indifference to the concept of reading. I don't recall any particular pre–first grade instruction from my mother and father. It apparently wasn't important to them that I begin first grade ahead of my classmates or be able to do anything that might convince the teacher I was special. In fact, Mom and Dad felt strongly that, outside of their parental eyes, none of us were special and should ever expect preferential treatment. "You can do anything you want, but your success is based on effort, not on some condition you were born with that makes you better than anyone else." Nobody in my first-grade class was particularly gifted, as far as I could tell.

I was a scrawny kid with mousy blond-brown hair tied back into a ponytail so tight my eyes "went Chinese." Every morning I stood on the flopped-down toilet seat for my turn at having Mom brush my hair with her latest buy from the Fuller Brush salesman. She would slick my hair back with brush strokes so hard that my head can still tingle at the memory. Then she pulled the wad of hair back as tight as tight would go into a ponytail she meant to last the whole day. She banded my hair with nothing like the coated rings they have now, but just a normal rubber band. My head stung like hell when the band came out at night, not to mention when she twisted and wrapped it tighter and tighter like a tourniquet onto the ponytail in the morning.

My only previous experience with organized academics was Sunday school. Mine happened in the basement of Zion Lutheran Church. Each Sunday one of our mothers read us a Bible story; then we cut out pictures of Mary and the Baby Jesus and pasted them onto a picture of a stable. Mary looked calm and serene, not at all as you would imagine someone would look after just giving birth to the King of Kings. Baby Jesus looked pretty much like any other six-month-old baby except for that golden light emanating from behind his head. For pasting we used that white glue made from horses' hooves that tasted sort of good, even though the Sunday school mother always told us not to eat it. While the glue dried underneath Mary and Jesus, we sang a rousing chorus of "Jesus Loves Me" and called it a session. Our teachers then exiled us upstairs to church, where we drew pictures on the communion cards or folded the church bulletins into rowboats until the minister finished the sermon. "May the Lord bless you and keep you. . ." would always snap us out of a self-imposed trance, the benediction signaling that we could retrieve our patent-leather slip-ons from wherever they lay underneath the pew and get ready to make for the car.

I found first grade a bit confusing. Gustie Zehm was my teacher. She was somehow related to Leonard, our neighbor, but I never knew quite how. Like many families in the area, the

Zehms had grown into many families overlapping and sharing the same last name. Gustie wore many hats: she had baby-sat us, was Darlene and Dennie Zehm's mother, had been both Marsha's and Cheryl's first-grade teacher, and belonged to our church. Gustie had been to our house innumerable times, often sitting out in the backyard by the lilac trees organizing a Women of the Church or Ladies Aide event with my mother. As soon as I entered first grade, I was no longer supposed to call her Gustie but, instead, "Mrs. Zehm." Not only that, but all of a sudden, Gustie, who had always been the picture of kindness, began to order me around, telling me where to sit, how to open a book, and mentioning fairly frequently that I must now be quiet.

I was quite a bit different from my sisters. I got the feeling early on that Gustie, who had also been their teacher, preferred them to me. Marsha was four years older than me. As the oldest, she had assumed a set of prudish qualities, perhaps considering it her duty to be as close a clone as possible to our mother. During her early school years she was always cheerful. She begged to be milk monitor in school, asked the teacher for permission to help pass out papers, and, whenever she didn't have to ride the school bus home, volunteered to stay after class to erase the blackboard. She was very popular. Her cheerfulness was a natural magnet for the other children, who didn't seem put off at all by her brown-nosing. It must have come across as sincere, and maybe it was.

Marsha resembled the Hein side of the family: large-boned, like Grandma Hein, and a little chubby, with long, dark-brown hair pulled back into one of Mom's killer ponytails. Marsha is still large-boned and a little chubby, although now her hair is streaked with gray and it's only medium length. She's still quite cheerful, although no one would call her prissy. Since her hormones kicked in, she has had one long parade of boyfriends and husbands, none of whom could be described as the least bit like Billy Graham or Mr. Rogers.

Two years Marsha's junior, Cheryl has always been very different from Marsha. Cheryl, like me, resembles the Keller side of the family. We're much smaller-boned, have long narrow faces, don't worry too much about our weight, and have large blue eyes. For most of her life Cheryl has had long, blond hair that stops at the small of her back. She was the serious, studious one of us, the one who did an excellent job on her homework, sat up straight all the time, and whose clothes were rarely torn or rumpled. She always won the spelling bees, got the math certificates, and gave the class speeches. While never volunteering for menial tasks such as hall duty or cleaning the erasers, if asked, she would graciously take on the job with the same dedication with which she approached her homework. She has continued to excel throughout her life, in spite of one major error in the marriage department, an error she has more than compensated for by raising alone two flawless sons.

Then I came along. I was surly as a kid, and still can be. I never liked helping the teachers and resented being asked. I enjoyed homework in reading and writing and slopped through anything else, with grades to prove it. The approval of my peers was moderately satisfying and the approval of adults was completely unnecessary. I think I have always considered myself slightly better than most, a consideration based on nothing concrete. I resembled Cheryl physically in our early years, and still do, except that my hair is brown and has been short since second grade, when Mom's ponytails started pulling out my hair on the sides. One day she hauled me, much to my horror, to the barber at Newberry's and had my auburn tresses clipped into a pixie cut. Adults thought the cut was cute, and I was furious.

As a result of who I was (surly) and who I wasn't (my older sisters), I think I was a bit of a disappointment to Gustie, whom I kept calling "Gustie" until she kept me after school one day. She told me I wouldn't be able to play in our classroom imitation grocery store until I quit calling her by her first name and quit talking aloud in class without permission.

I loved the imitation grocery store Gustie set up in the corner of the classroom. It consisted of a tiny cash register and empty containers of Frosted Flakes, Cracker Jacks, and Campbell's Soup. We children delighted in taking turns buying empty boxes from each other using Monopoly money. Apparently Gustie was a pioneer in the area of "hands-on learning," using this method to teach us how to make change, how to add numbers, and, ultimately, how to make a buck.

The rest of first grade I hated. For hours, it seemed, we had to sit in utter silence, to what purpose I did not know, as Gustie, who had become to me not my mother's friend but my boss, droned on and on about the b-b-b-baby sound or why our "t"s should be crossed slightly above the dotted blue line on our paper. My first-grade report card gave me a bland "Satisfactory" in every category and under the section for comments noted: "Teri is a bright little girl who, if she applied herself, could do excellent schoolwork. However, she spends too much time talking during class and does not take pride in her work." It seemed peculiar to me that Gustie would write these official pronouncements about me rather than just tell Mom over coffee when the two got together. Besides, who was she to tell Mom what I was like, as if Mom didn't already know?

* * * * *

My mother started school before harvest was over, a couple of weeks before me. In deference to her education, my father hired an extra worker to drive the truck she usually drove during the wheat season. Dad reduced her farm duties to keeping us children in line and making the meals. I'm sure she actually reduced herself, since Mom has never been one to let other people reduce her to anything or dictate how she spends her time. These were the days before Minute Rice and microwaves. Mom prepared three big meals every day for four children, herself, and three or four ravenously hungry men. The tuna casserole, green salad, and chocolate-chip cookies were all homemade. We grew the carrots in our garden and bought the ice cream at the store. That was lunch. Mom had the dinner roast, gravy, mashed potatoes, green peas, and apple pie in the oven on warm while she got into her dress clothes and put on her face for class that night. At ten years old, Marsha knew how to set the table and to put down the hot pads before pulling the food from the oven when the men came in for dinner. It was dark by the time they had all showered and come upstairs. We girls had already eaten, put on our pajamas and stationed ourselves in front of the television. If we had a bath that night, it was long over, so that the hot water tank could fill again for the men. Tracy was fast asleep by the time the men sat down for dinner, Cheryl and I having taken turns bouncing her around as she screamed her nightly rage beginning the moment Mom dashed out the door.

Mom drove to night school two or three times a week for four years to get her bachelor's degree in elementary education. By then she was so accustomed to the drive that she kept driving for another four years to get her master's degree.

In the summers she dropped her daughters off at the YWCA in Spokane to take swimming lessons while she sat in the city library across the street to work on papers. Mom was always working on papers, which she clacked out on her black upright Smith Corona on hot afternoons between pies or in the evenings when she didn't go to class. From that Smith Corona came such academic works as "From the Celestial to the Snake Pit" for her Milton class and, less ethereal, "Trends in Teaching Reading" for Curriculum 300.

Those college papers propelled us into becoming very good swimmers. We crawled and backstroked our way through Beginners I and II, Intermediate I and II, and, ultimately, Swimmers, a rigorous course that required the recipients of the certificate to swim ten full laps of the pool, alternately using the crawl, the breast stroke, and, for one torturous lap, the butterfly. After Swimmers, I retired. Enough was enough. Marsha, however, went on to higher glory by passing Life Savers, even though none of us would let her practice mouth-to-mouth on us.

While Mom was off educating herself we were doing the same. Tracy may still have the scars from the pins jammed into her baby butt during those nights of readying her for bed. To this day there is a charred mark where Marsha roasted the kitchen counter with the broiler pan, dropping it there when her finger found the hole in the hot pad. We learned that the toilet overflows if you try to flush down a whole helping of the Brussels sprouts you don't want discovered later in the waste basket. We found that if you accidentally turn off the oven while the roast is cooking, you cannot retrieve the lost time by turning the oven on the highest setting for the ten minutes until dinner.

* * * * *

This time of learning was equally shared by the Brewer girls, who lived over on Hays Road, probably a mile and a half away, as the crow flies. Pat was the oldest Brewer girl, although she wasn't exactly a Brewer, because her real father, Don, had died because of the war. He was only twenty-one and had been home from the war for only three months when, lying on the couch one day playing with his toddler daughter, a piece of shrapnel left from a war wound, a piece he didn't even know was there, worked its way over to his heart. Just like that, right there on the couch, Don was dead.

Then Harriett, Pat's mother and Don's widow, met and married Ed Brewer. Born and bred in Spokane, Harriett considered herself a city girl. She was shocked to find herself dating a farmer, even more shocked to fall in love with him, and, finally, completely shocked to consider a move to a farm in what was, to her, the middle of nowhere. The old farmhouse Ed showed her was a wreck with peeling oilcloth on the walls and a large hole by the stove. Ed promised to fix it. She agreed to move to the farm, but only after extracting from Ed the promise that they would return to the city when Ed retired at age sixty-five. "This is for love, and nothing more," she emphatically told him. Harriett was twenty-eight years old then and lived on that farm for over thirty years before she finally got to return to the city. Ed's retirement came a year earlier than he expected, at sixty-four, because he was diagnosed with cancer. He wished to have Harriett installed in the city home she had been yearning for all those years just in case the cancer killed him.

Ed and Harriett's first child was Mary, who was born the same year as Marsha. Their second child was David, born eighteen months later. Brenda was their last child, born near Cheryl's birthday. My parents went on to have two more children, but the Brewers stopped with Brenda.

Mary and Brenda, due to their proximity in age to Marsha and Cheryl, were frequent visitors around our house. Marsha and Mary were very similar in personality and appearances. Both were kind of chubby, and both were happy most of time. Mary was loud and brusque and a perfect target for my dad's teasing. On the other hand, Brenda, like Cheryl, was much more serious and graceful. She was tall and thin, had perfect posture, and possessed this melodious, gentle laugh. I remember feeling pudgy and uncouth around her.

When Mom went off to school and handed over more and more household chores to her daughters, Harriett decided it was time for her daughters also to take on more responsibility. The two housewives hatched a plan to put the oldest girls, who were now teenagers, in charge of the kitchen for two weeks in the middle of harvest. The girls had to plan, shop, and cook all three meals a day for the hungry harvest crews.

Mom and Harriett's motives were slightly different. Harriett thought it was time that Mary began to learn the responsibilities and skills necessary to be a good wife. How to feed a crew of hungry men was an essential part of being a good spouse. Mom, trying to thwart Marsha's inimitable romantic spirit, wanted to give her a taste of reality. By then, Marsha had blasted into adolescence and was on her third "serious" boyfriend. This time it was Dick, our hired man and neighbor boy, who was an older man at eighteen, and already a high-school graduate. Marsha definitely had been entertaining fantasies of that magic moment when she and Dick would march down the wedding aisle and be pronounced man and wife. They would get all those presents, go on a honeymoon to California, and get their picture in the *Standard Register*, looking radiant and uncontrollably in love. Cheryl and I discussed periodically how difficult it was to conjure up the image of Dick being radiant or uncontrollably in love. He was incredibly shy and rarely showed any emotion other than embarrassment. The closest we could get to imagining him radiant was picturing him sunburned, which he usually was since he was so fair-skinned. Marsha, on the other hand, was very easy to imagine uncontrollable.

I don't know how much to attribute Harriett and Mom's machinations to the girls' eventual lives, but today Mary is married to a farmer and lives in Moscow, a Palouse wheat town on the Idaho side of the border, and Marsha is a nurse in California, living alone. Marsha is, in fact, an excellent cook, thanks to Mom's instructions on pot roast, and her first husband, an Italian doctor, who taught her the magic of pasta sauces. Dick didn't survive long as a boyfriend after the experimental cooking weeks, if I remember correctly.

Mary and Marsha were completely gungho about their new, adult, wifelike responsibilities. They loved to page through recipe books, cook things that took all afternoon, and turn the kitchen completely topsy-turvy in the process. Both accepted the challenge with their characteristic enthusiasm. For the week before the cooking began they were on the phone several times a day to each other, in heavy consultation regarding recipes and cooking times. They decided to prepare identical menus for their particular harvest crews. They also decided it was their responsibility to give their harvest crews culinary experiences that perhaps they had previously not had.

The girls had matching copies of Dorothy Dean's cookbook *Homemakers Service,* a subscription recipe program that many of the housewives in the area participated in. Every month, a two-sided flyer would arrive, complete with holes punched to fit into the complimentary notebook that came with each subscription. The flyer was filled with recipes, all fitting into some declared theme, such as "Party Time Pleasers" or "Calorie Countdown."

The girls launched their first cooking week with Dorothy's "Hawaiian Hospitality Night," an unusual meal choice for a group of harvest-hungry men who were expecting meat and potatoes. First off, the girls decided to give everyone Hawaiian names, as Dorothy's flyer conveniently listed them with their English equivalents. Charles was Kale, David was Kawiki, Margaret was Makaleka, and Mary was Malia. Unfortunately, Dorothy's list was far from complete and didn't supply the Hawaiian names for Ralph, Dwight, Dick, Russell, or even Dolores. The girls didn't care, and happily set place cards in front of each plate. "Tonight," Marsha announced to Dick when he came upstairs from the shower, "you are Paolo. I am Ianete." He looked embarrassed as usual.

The menu that night consisted of Polynesian Pork, which was pork cooked in pineapple juice and vinegar, along with steamed rice, and Oahu Stuffed Banana, an interesting dish in which the girls mixed baked banana pulp with cheddar cheese and butter, restuffed the banana skin, baked the mixture again, and served it to the harvest crew. "What the heck is this?" asked Lopaka, otherwise known as my dad. The Fluffy Yam Soufflé was a little underfluffed, and the Halakahiki Yogurt Bread was "substantial," as Setepano, a.k.a. Dwight, pointed out. "I think the dates and pineapple yogurt really make this recipe," he said graciously. Marsha, or Ianete, had dessert to top off the meal. My dad gave the crew a stern look when Marsha announced their dessert treat for the evening, naming it in her newly acquired Hawaiian accent, sort of a monosyllabic cavewoman dialect of clipped-off words that she imagined the Hawaiians spoke. "And now, we have Ono Coconut Cake." It wasn't half bad, in spite of the name, at least compared to the Oahu Stuffed Bananas, which even Paolo, who was deeply enamored with the chef, couldn't eat.

That dinner was just the start. There were bigger things to come. The girls were definitely inclined toward recipes that sounded foreign: Turkey Tetrazzini, Chicken Coq au Vin, Tuna Noodle Florentine. A lack of ambition was not their problem. They tried Chicken Artichoke Mousse, Seafood à la Ritz, and even Ham-Broccoli Fondue one night. Desserts were always a new experience: Maraschino Cherry Cake, Frosty Grasshopper Torte, and Wacky Lemon Custard.

For the most part, their attempts were not particularly successful, undoubtedly due to a disparity between their ambitions and their abilities. They both attempted a baked Alaska, although the ice cream melted all over both of their ovens before the meringue was barely warm. Their béarnaise sauces curdled at the same moment, and their ladyfingers were overbaked and dry. Mary forgot to add sugar to her rhubarb pie, and Marsha's potato rolls refused to rise. The lattices on the cherry pies burned to a crispy black, and Marsha overcooked her baked herb carrots to an orange, pulpy goo. However, the two girls determined to persevere.

In the mornings they jumped out of bed before dawn, called each other for a wake-up, and dove into such breakfast treats as eggs Benedict and soufflés, or, more accurately described, soufflé attempts, since for three breakfasts in a row, Marsha's Never-Fail Cheese Soufflé resembled a yellowish soccer ball with the air half let out. On those mornings she made a quick substitution of Frosted Flakes, with raspberries she'd picked from the garden the day before. She took pride in her ability to make a quick decision in the likely event of failure.

Marsha took to putting a little garnish of parsley on top of everything she made, whether it was breakfast, lunch, or dinner, hoping to add color or create a distraction from any meal that resembled a deflated soccer ball. Mary's garnish of choice was curled carrot ribbons or her version of radish flowers, which her brother, David, suggested she instead call "cut-up radish pieces." Mary had even ordered a special set of curling knives from a late-night television ad, sending a check to Wisconsin and paying extra for a rush delivery. Unfortunately, she was too exhausted every evening to spend the time necessary to get the cuts precise enough to transform that simple radish into a blooming rose.

We learned later that the hired men from each farm would compare notes before and after meals. They speculated about whether dinner would be recognizable and evaluated, after each eating event, whether it would be considered edible by the general population. It was agreed that Marsha could be counted on for a more agreeable casserole, but that Mary's piecrusts were better. "Well, well, what have we here?" Dad would graciously ask, as Marsha lowered a platter of food onto the center of the supper table. "It's Chicken Parisienne," or "Hamburger Fiesta," she would announce.

At first she said it proudly or with a giggle of delight that demonstrated how happy she felt with her masterpiece, or at least the intrepid spirit that drove her to attempt it. As time went on, though, and the soccer-ball meals deflated more and more, Marsha, who wasn't accustomed to getting up before nine in the summers, and who had been chained to the kitchen since this project began, started getting a little testy in her responses. "It's Green Tomato Pie, of course," she would say defensively, just daring anyone to imply that the dish didn't sound so good and, in fact, looked a little dubious to boot.

About midway through the second week, both Mary and Marsha overslept and missed breakfast one morning. This was the beginning of the end. Breakfasts became cold cereal and canned fruit. Lunches metamorphosed from French onion soup, German potato salad, and homemade whole-grain rolls to tuna sandwiches made with Wonder Bread and a side of potato chips. Dinners were taken from Section X-24 of Dorothy's book: "Can-Opener Cuisine." The girls chose recipes such as Can-Opener Clam Casserole and Hasty-Tasty Hash, along with a head of iceberg lettuce and a bottle of French dressing for salad, slapped down on the table. By then ice cream was always dessert. On the very last night, Marsha got a bit of a headache and asked Cheryl to put out the food, which was, on this grand finale evening, a stack of Swanson TV dinners.

The experiment was a great success. Marsha decided she was a bit hasty to consider becoming Dick's wife and set her hopes on a career that enabled her to hire a cook. Mary decided if she was going to be a farmer's wife, she had better start collecting recipes for casseroles that could be prepared and frozen way ahead of time. The hired men learned that asking Dad for hardship pay while his eldest daughter was doing the cooking was in no way an insult to Dad, even though he refused. On the other hand, Ed Brewer gave each of his crew a twenty-dollar bonus for each of the two weeks, and everybody knew why, even though Ed wouldn't say. Cheryl and I, as well as Brenda and David, learned what a baked Alaska and a soufflé weren't, and Mom and Harriett, both women with great senses of humor, got a million laughs over those weeks, especially on the last night. While the men consumed those piles of TV dinners, both girls lay prone in their beds, never again to attempt a Lazy-Day Chicken Divan.

Growing up in the fifties meant we feared at any moment that we could be bombed to smithereens by the Russians. Did we in the Inland Empire feel this possibility more strongly than others in America because we had Hanford, our own personal atomic bomb factory, nearby? Who knows? I do know that, like all kids growing up then, we knew exactly what to do when the air-raid sirens went off, thanks to the drills.

I think it was more like the early sixties when we learned the drills, because I remember them from under my second-grade desk, which was in 1960. Mr. Evans, our grade-school principal, came in and said, "We have work to do," in a voice so solemn we felt, for a brief moment, ageless. Mr. Evans looked like an overweight Buddy Holly, complete with dark, slicked-back hair and black, horn-rimmed glasses. Mrs. Miller, who was probably twenty years his senior, solemnly gave him the floor, knowing that the business at hand was serious, necessary, and probably more appropriately conducted by a man.

He didn't mean to scare us, he said, but we all knew that the Russians were trying to take over the world and that America, as the best place God had ever created, was the biggest obstacle for them. Even though Good was winning the fight over Evil, the Russians weren't beaten yet and could be expected to do anything, even to innocent children. Because of this, we had to get ready just in case the Russians attacked. The best way to ready ourselves was to practice having a bomb dropped on us.

First, Mr. Evans told us, we would hear the air-raid siren, an alarm that had been installed at the post office downtown, two blocks away. In fact, once he had made it to all the classrooms, Mr. Evans was going to call Glen, the postmaster, to activate the siren for a bomb drill. The whole town knew about the siren, which went off periodically from then on whenever we had our practice drills, not to mention once or twice more when it malfunctioned and nearly gave heart attacks to a couple of the elderly at the nursing home.

That day in class, Mr. Evans said if we heard the siren while we were at school, we should drop down, hide under our desks, and wait for further instructions from Mrs. Miller. With this, he turned and marched out of the room, presumably to inform the third graders.

I remember that first drill. Mrs. Miller tried to re-interest us in our subtraction lesson, but it was hopeless. We were waiting to practice being bombed. Finally, we heard the *whoop-whoop-whoop* of the siren, looping through the air from downtown. We dove under our desks, knocking papers and Peechee folders everywhere. The wooden floor was dirty, splintery, and rutted in a way I'd never noticed before. It was hard to compress my body enough to fit completely under my desk. Every time I moved, a dress fold or my shoulder would edge out across the invisible boundary. Cory Sunkle, under the desk next to me, kept whispering, "I . . . am . . . a . . . Russian, . . . " and then making bomb blow-up sounds as best he could in a whisper. He lay on his back with his legs extended up, balancing the desk over his body, shifting it from one leg to the other, while making the noises. I hated sitting next to him that year, and I hated hiding under my desk next to him hiding under his. He was one of those kids who would probably have been on Ritalin today.

As we scrunched under our desks, trying desperately to keep our clothing in line, Mrs. Miller paced in front of the group, snapping at children whose extremities weren't neatly tucked enough into their bodies. Finally she ordered us to line up in silence and follow her single file down to the school cellar door, where we were joined by equally silent lines of somber children. None of us had ever been inside the cellar, which wasn't really down any farther than we, on this basement floor, already were. The cellar was a room off the cafeteria that Mr. Kamer, the school janitor, used as a storage closet. The school water heater was there, as well as a giant furnace, lots of pipes, and a miscellany of other things.

On our drills, of which there were several, we never actually went into this room. There really wasn't enough space for all the kids, but it was our understanding that if there ever was a real nuclear attack, Mr. Kamer's workbench and the other janitorial stuff that crowded the room would be removed to make way for us. Also, in the event of a real disaster, the room would be miraculously stocked with enough saltines and Kool-Aid to keep this school full of kids alive for the years it would take for the nuclear fallout to drift away.

I thought the whole thing sounded great. I liked saltines and assumed this meant we wouldn't have any schoolwork to do. I had greatly enjoyed the few times I'd been allowed to stay overnight with my girlfriends, so this seemed like a huge dose of fun, even though I knew I would miss my parents.

Now that I think about it, I don't really know what kneeling under our desks was supposed to accomplish, except, perhaps, to keep us busy while the grownups filled the shelter with saltines. If I had been Mrs. Miller, I would have used the time to run for my car. Now, as an adult and a teacher, I would rather die in a nuclear holocaust than spend more than a night in an elementary school cellar with a mass of Kool-Aided-out kids.

Back then I never seriously thought the nuclear bomb attack would happen while I was at school. I knew I would be a lot safer if it happened while I was on the farm under the watchful eye of Mom and Dad. I begged Dad to build a fallout shelter, not because I was terrified of what would happen to us if we didn't have one, but because those little secret rooms dug into the ground seemed so mysterious and wonderful. I imagined the air-raid sirens going off, apparently assuming we could hear the one all the way from Waverly. I pictured our whole family racing toward the shelter, each girl with a stuffed animal under her arm as she made for the ladder leading down into the ground. Dad would shove the livestock into the barns, pile their mangers with a lifetime supply of hay, and grab our favorite small pets to take down with us. Mom, who had already stocked the shelter with all the food and drinks we would need for the rest of eternity, would grab—well, I didn't know for sure what Mom would grab. Maybe the radio or her typewriter. Anyway, we all would climb down into the shelter, Dad last, and he, with a resounding boom, would lower the trapdoor just as the nuclear destruction began to occur. Of course, we were completely safe in our fallout shelter, which was cozy and warm and felt just like home. It was kind of like being part of the Anne Frank family, only nobody evil was looking for us. I imagined the fun we would have down there as a family, playing Yahtzee and Clue. Night after night we would play by the light of kerosene lamps as we listened to the radio, keeping track as the forces of Good triumphed over Evil, not to mention Miss Plum in the conservatory with her lead pipe.

Unfortunately, Dad refused to build a bomb shelter. I have the feeling now he thought it was pure folly to think we could survive anything like what they were describing on the television.

I remember Mom telling me that if there were a nuclear disaster we would put burlap bags over the tiny basement windows to keep out the nuclear poison and live down there with its wood stove, freezer, and bathroom. The basement has, of course, lots more room than the fallout shelter of my dreams, and if Mom said burlap bags over the windows would keep us safe from toxic nuclear waste, it must have been true. I was, however, still disappointed.

I was also concerned for the neighbors, knowing that the Zehms, the Brewers, and the Dennies didn't have basements or fallout shelters. Mom assured me that they all were welcome to join us, and I was left to contemplate our basement full of the seven Zehms, seven Dennies, six Brewers, and six Heins. We were going to have to stock up on Yahtzee score pads for sure.

Then I hatched the idea of sending the Brewers over to the Hahners. This was going to be very disappointing to Cheryl and Marsha, who would really enjoy Brenda and Mary's company in our fallout basement, but we simply did not have enough supplies to have all the neighbors over, in spite of how big our basement was. The Hahners lived very close to the Brewers, and there was a sweet little friendship between nine-year-old David Brewer and five-year-old Greg Hahner. It would mean a lot to Greg to have David so near for such a long time, as I imagined our residency in the fallout basements would be.

David was Greg's hero. Greg was the oldest in his family, so, in his five-year-old way, he felt somewhat desperate for an older brother. David, surrounded by sisters, was happy to take on this surrogate little brother who lived just a mile down Kentuck Trail Road. Luckily for Greg, there were no other boys near David's age on our bus route, so the elder was happy to sit next to Greg every morning on the school bus in the seat that Greg, who was one stop ahead of David, claimed for them both. One would have thought that Greg would rather have sat with his cousin, Bruce

Dennie, who was the same age and got on the bus one stop before Greg, especially because Greg was very shy and should have cowered in the presence of the fourth-grade David. But David clearly had a soft spot in his heart for the kindergartener and was even known to chase kids out of their seats in order to share one with Greg. I imagine that David enjoyed the hero worship.

The Hahners had a small basement, but one I was sure would fit both their family and the Brewers. The Dennies probably also should have gone to the Hahners, as Claudia Dennie, my friend Gail's mother, and Jimmy Hahner were brother and sister; but this was my plan, and I wanted Gail Dennie, my good friend, in our fallout basement. Otherwise we would be stuck with all the Zehm boys, undiluted, and they were terrors to spend so much time with unassisted by a girlfriend.

* * * * *

That year, 1960, the Fairfield Service Club chose "nuclear war preparedness" as their theme for the Flag Day float. The service club always had a community service theme to their floats, which usually made them a little boring. That year they built a replica of a fallout shelter on Bill Osteller's hay trailer. The shelter was sort of cut in half, so we could all see inside. On the makeshift shelves were lanterns, a radio, boxes of crackers, and a Bible. Their fallout shelter seemed a little short on board games, I noticed. Two giant banners that read "Be Prepared" were hung on either side of the trailer. On a bench in this pseudo fallout shelter sat Donna Olson as mother, Carl Robinson as father, and Ricky Loeffler and Susie Groth as the children. They were to represent a happy, secure family, all safe and snug in their fallout shelter, while the nuclear war raged outside. This phony family did look happy, except for Ricky, who had been volunteered for the role by his father without the son's knowledge. There was a rumor in our class that Susie had a crush on Ricky, a source of great embarrassment for him. He was of that age, eight, when those sorts of possibilities are highly embarrassing. Not knowing how to digest this threat of affection, and not even knowing for certain whether it was true, Ricky had decided to take the prudent route and not show any friendliness toward Susie whatsoever. Not too long after making this decision, he found himself on this float with her, parading down Main Street in front of everyone, including his friends who had told him she had the crush. The nuclear age was destroying Rickie's private life. Where he really wanted to be was on the street, in line with his friends to catch taffies thrown from other floats, not trapped in a bomb shelter with Susie Groth.

The Flag Day parade is to Fairfield what the Macy's Thanksgiving Day parade is to New York City. It happens once a year on the Saturday closest to June 14 and is bigger than life in the minds and hearts of all the children who have lived in the area during the past eighty years.

To be honest, I don't recall anyone ever mentioning the flag, the nation, or patriotism during any of the Flag Day celebrations that I attended, but I guess some things are just a given. I imagine that's how Harry, the town mayor, would have explained it to us. As mayor, he was the king of Flag Day, officiating at the awards ceremonies, riding in the parade convertible, chairing planning-committee meetings, and judging the Kiddy Parade contestants. On this one day each year, hay trailers came to a halt, farmers put machinery repairs on hold, everyone postponed house and barn cleaning, and people flocked to town.

The centerpiece, as well as kick-off event, of the day was the parade. Anyone could be in the parade, which started at the top of the hill by the grade school, wound down Main Street, over the tracks by the grain elevators, up past the park, and turned to the right after the six blocks of Main. At that point the parade was over. It probably took thirty minutes, maybe forty, to traverse the entire parade route.

Children began their Flag Day careers in the Kiddy Parade. Any civic-minded parent was obliged to fashion clever costumes to adorn their children for the walk. Preferably entries were something that would integrate either a pet or a younger sibling into the theme. We won the blue ribbon in 1960, the year we transformed our red wagon into a throne with a combination of chicken wire stuffed with pink Kleenexes and Marsha's pink bedspread. Cheryl pulled the wagon, dressed as a court jester, while Tracy, our toddler sister, dressed as a queen and sat in the wagon, holding on to our toy poodle, Fifi's Jacqueline. The dog had a small foil crown tied around her head, as did Tracy. Dressed as a jester like Cheryl, I walked behind the wagon carrying a sign that read "Guess Who's the Queen in Our House?" Our minifloat had it all—the baby sister, the cute dog, good costumes, and a chance for audience participation. I didn't much like our parade contribution that year, however, because I wasn't the one the sign was referring to, although I did perk up when we won the ribbon. Maybe Mom asked Harry to slip us a blue so I would quit pouting.

The year before, Mom, who always designed our costumes, honored the Native American community by sending Cheryl and me out onto Main Street as a teepee and Chief Qualchan's wife, the Indian princess, Whiet-alks. This was shortly after Mom read at the Spokane Library the story of Colonel Wright's hanging of Chief Qualchan. Mom wrote a long paper on it for her Pacific Northwest history class, in which she resoundingly took the side of our Native American forefathers who, she posited, had come strictly in peace to the river, only to be ruthlessly strung up by the colonel's men. She speculated to us, her daughters, what it was like for Whiet-alks to witness the hanging of her own husband and then to have the courage and strength to drive that spear into the ground and ride off. We were left with a romantic and tragic image after the story weaving of our mother, and I, for one, was completely ready to honor Whiet-alks in my Flag Day representation of her.

This time it was Teepee Cheryl who was the unhappy one. Mom had fashioned a cone from chicken wire and wrapped cut-up burlap grain bags around it. On the burlap we had painted with tempera what we imagined were Indian symbols. Inside, Cheryl wore an old straw hat that supported the top of the cone. Mom stuck empty wrapping-paper cardboard tubes up through the top of the cone to represent the poles coming out of the top. I thought the whole thing looked extremely authentic. Cheryl, unfortunately, hadn't a clue as to how it looked because she could barely see a thing, not to mention that she was cooking inside this, her own personal Indian sweat lodge. Another problem for her was walking. Unless she moved at a snail's pace, Cheryl risked upsetting the delicate balance between teepee and straw hat, although with her limited vision and chicken-wired costume, gazelle movements were unlikely. Mom had cut a small hole on the right side of the costume where Cheryl could extend a hand, which I was to take and, hence, pull her along. Neither one of us, however, wanted to hold hands in public. We decided that Cheryl, who could, if she looked down just right, see a bit out of the opening, would watch my moccasined feet as we went down Main Street. For six blocks we could pull it off, Mom reasoned.

On every farm there is an abundance of burlap grain bags, so just about all the Flag Day costumes were in some way or another created from a former oat or seed sack. My Whiet-alks costume was hot and scratchy, but I could see fine and I liked the headband with the magpie feather stuck in the back. I had a new pair of moccasins from Woolworth's and felt as close to a real Indian as my German Lutheran self ever had or ever would. I had a great time as that Indian princess, practicing my "woo-woo-woos" and doing little rain dances down the street. I suspected that rain dancing and woo-wooing were more characteristic of Indian brave behavior, but I could think of no alternative.

I found it hard to do a rain dance and stay close enough for Cheryl to see my feet. After the first block I began to dance away for brief moments, retreating back to Cheryl before she could lose her way. About halfway through the parade, right in front of the bank, I spotted Linda Peterson up ahead, parading down the street dressed as Tinkerbell. This Flag Day was soon after Mary Martin had first played Peter Pan on Broadway and the Spokane Civic Theater had created a facsimile. People just couldn't get enough of that play, especially Linda, who had seen it three times.

On this Flag Day, Tinkerbell's costume designer was Linda's older sister, Pammie, who loaned Linda an old ballet tutu and had lathered her third-grade face with liquid face makeup, eyeliner, and enough powder to clog up a small dam. Pammie wanted to be a cosmetologist when she grew up.

I had been looking for Linda before the parade started, so when I saw her, I jumped up ahead to say hello, leaving Cheryl, the lone, blind teepee, to hobble down the street without me. I meant to stay for only a minute, but the excitement of the moment, plus the satisfaction of my whooping to Linda's Tinkerbell twirls, was so much fun that it proved a distraction for the two of us.

When it became evident that I wasn't returning immediately, Cheryl, sweating and stumbling her way down Main Street, could have thrown in the towel, or the teepee, as the case may be, but she didn't. Cheryl has never been a quitter. She was in the middle of the Flag Day parade and, while she couldn't see through her burlap haze, she imagined huge crowds, all eyes trained on her. She continued to wobble down the street in slow-motion baby steps, a sad symbol of our treatment of the Indian Nation and, perhaps, in some abstract way, an even greater representation of the Chief Qualchan tragedy.

Mom could have killed me, seeing Cheryl's dilemma from her lookout by the park. She sent Lynn Olson, who had already finished the parade dressed as a tabby cat, back into the street to rescue Cheryl. Together, teepee and tabby cat finished the parade just a block behind Tinkerbell and Whiet-alks. I'm certain I got in big trouble for abandoning Cheryl. All I remember about that time was being quite upset when Grandpa and Grandma Hein, who always came down for Flag Day, and had witnessed Cheryl's struggle from their folding chairs in front of the Pea Growers, limited the number of quarters they gave me for rides that year. They said it was to show support for Cheryl and her teepee trauma.

Immediately following the Kiddy Parade was the real parade. We knew just about everybody who participated: those who were riding their horses, playing in the band, marching with the drill team, or just waving from a car. The best part of the parade was the floats. Each had at least one pretty high-school girl in her prom dress, waving regally with poised hands in elbow-length white gloves.

Next came the baseball game. When I was young my dad was often the announcer, bellowing out play-by-plays while his buddies ran the bases. Fairfield's team was known as just "the town team." We didn't need to designate the sport, since it was the only team we had, and we didn't need the name of some wild animal to describe our better qualities. We didn't need to be called the Fairfield Tigers to let everyone know our guys were fast and mean.

Dad played on the team for years, usually third base, and quit only after his thyroid surgery, when he started to limp. The team played about ten games a year and were what you call "semi-professional." They charged the spectators fifty cents each and paid the pitcher, who came down from Spokane, ten dollars a game. Everyone else was from Fairfield and happy to play for free.

That's how Dad and Milo Gorton became such good friends. Years before I was born, the Fairfield team saw Milo playing for the Spokane Valley team and convinced him to pitch for Fairfield. Milo was a teacher in Spokane at Bemiss Elementary, and I think he must have rather liked getting away from the city because he pitched in Fairfield for several years running. On game days his wife, Marge, came down with him to root along with the other wives.

After the baseball years, the Gortons still came to visit, along with their children, Larry and Susan, who were, among other things, deaf. They were only slightly older than us and very good-looking. My sisters and I practiced speaking loudly and slowly around them, enunciating clearly, as both Larry and Susan could lip-read almost as well as we could hear. We Hein girls found Larry and Susan's stone-deafness only a slight inconvenience, if not sort of intriguing. We were especially impressed with lip reading, something we would practice with each other for hours after each Gorton visit. "*Whaa a a t t t t a a a a m m m IIIII sa a a ay ing?*" we would pose to each other with our newly elasticized lips.

Not only did we envy Susan and Larry their lip-reading ability and their good looks, but also that they got to go to Camp Easter Seal. At Sunday school the teachers were always telling us our offerings went to support Camp Easter Seal, a place on Lake Coeur d'Alene where handicapped children could have fun together. The teachers really talked this place up, presumably to make us all feel glad about turning over our dimes for those less fortunate. The Camp Easter Seal handicapped kids had ponies to ride, canoes to paddle, and even water-skiing lessons. At Camp Lutherhaven, where we unhandicapped kids went, we had two rowboats that leaked, a baseball field, and craft classes where we made things out of pine cones. The food was bad and dessert was Jell-O almost every night. Our Sunday school teachers made

Camp Easter Seal seem so exciting that it was hard to think of any kids who got to go there as truly unfortunate. Susan and Larry had pictures of themselves on water skis, something no one at Camp Lutherhaven ever, in a million years, got to try.

Dad and Milo quit the baseball team at about the same time—Dad for health reasons and Milo because the frequent drive was difficult after he had a family. But before and after their playing years, the Flag Day game was always the biggest of the season. Fans packed the stands on both sides of the bleachers, happy to watch the local guys match skills with Rockford, Tekoa, Rosalia, Mica Peak, or maybe even the Spokane Valley team.

Those uninterested in baseball could go to the Grange Hall and get a fried-chicken dinner prepared by the Women of the Church. That's where many old people whiled away the afternoon. My grandparents, Philip and Marie, handed us a pile of quarters and sat for the rest of the afternoon in the cool hall, passing the time with old neighbors and acquaintances who looked forward to this yearly reunion. We always knew where to look when the quarter supply ran out.

My parents were good for quarters, too, although, as the next generation, they were much too busy to monitor our financial needs. Not only did Dad announce the first half of the baseball game, but he also ran the ring toss when Carl Robinson went off to be in the talent show. Kids liked it when Dad ran the ring toss. He was happy to give them tossing advice as they handed over their red tickets for a chance at a kewpie doll. It was not uncommon for him to let them have more turns than their red tickets officially allowed, sometimes holding an empty pop bottle away from the rest to make ringing it easier. He would even chase that flailing ring through the air with the bottle, if that's what it took to make a winner out of a particularly pathetic contestant. Dad's ring toss never made much money for the Service Club, but it went a long way toward public relations.

Mom was plenty busy herself. She was often the pianist for the talent show, and she helped with the Ladies Aide booth in the gymnasium, selling preserves and embroidered kitchen towels throughout the afternoon. Every now and again she ran the Cake Walk, which raised money for Women of the Church. Mom never broke the Cake Walk regulations, especially for a daughter, and always ruled against me, even when the back of my leg was still touching the winning chair when the music stopped. I tried to play when Mom ran the Cake Walk with Mona Zehm, Leonard's wife, because Mona could usually be counted on to stick up for me. "I'm pretty sure her leg was still touching the chair, Dolores," Mona would say with a wink in my direction. Other times, Mom ran the Cake Walk with Harriett Brewer, who was also wonderfully lax about the rules. We were disappointed when Harriett vowed never to do the Cake Walk again after the time she knifed her daughter Mary by mistake. It was a human error and not worth quitting over, but Harriett felt terrible. Mary was stealthily snitching some frosting with her index finger from a chocolate cake near the edge of the table. Spying Mary out of the corner of her eye, Harriett leaned forward and, without thinking, gave Mary a slap across her hand to stop the snitch. Unfortunately, Harriett forgot she was holding onto a knife she had been using to cut brownie squares. Mary screamed as the guilty snitching finger began to spurt blood. Several stitches later Mary was fine, but Harriett never recovered enough to assist in running the Cake Walk again.

My parents were both at their best on Flag Day. It didn't even seem to bother Dad that much that he couldn't play baseball anymore. Farming is such quiet, solitary work that I think he must have relished this one day of civic responsibility, hamming it up over the loudspeakers during the baseball game and driving float trucks slowly down Main Street while throwing candy from the cab.

I think Flag Day did and still does help people forget their troubles. It helps people forget about dropping wheat prices, the weather, whichever cow is afflicted with anthrax, and ailing elders. I think it helped my dad forget that damn limp, the one that made him quit baseball in the first place, not to mention hunting, and made everything so much more difficult. Even after other people started getting sick in our neighborhood—Mona with her brain tumor, Jimmy Hahner with his cancer, and the others—we never stopped going to Flag Day.

How could anybody be sad on Flag Day? It was so much fun. When not cake walking and ring tossing, we children rode the swings, the only real ride, except for a fairly pathetic merry-go-round. The swings consisted of wooden chairs, hung on chains extended down from a rather flimsy flagpole, and were propelled by a noisy engine. We would jump into a seat, fasten a thin little chain across us, and grip tightly as the ring of chairs began to turn. We cranked our heads sideways to grin tensely at our girlfriends, chained into their own chairs behind us. Slowly the engine started the chairs turning, picking up speed with each revolution, faster and faster until finally we were sailing through the air, holding on for dear life, our screams left hanging in the air for the swingers behind us to hear. Around and around we went, our heads turned sharply against the direction we were moving, the wind plastered against our cheeks, blurring our vision, and sucking out our breath every time we tried to look in the other direction. We just kept screaming. It was horrible, in a way, and our idea of an infinitely great time.

We most admired David and Brenda Brewer for their swinging prowess. They would find two swings one behind the other and claim them, each grabbing on to the other one's swing before the engine started moving the swing chairs in a circle. As the swings twirled higher and higher, David and Brenda would continue to hold on to each other's chairs, keeping them hooked together with the crooks of their elbows, until right when the chairs were twirling their fastest and highest, David would yell, "One, two, three!" and the two would let go of each other's swing, sending them both recklessly careening up and down while continuing to go round and round. We all screamed with them, even those of us still on the ground, and felt brave for even daring to watch.

David couldn't have been more than eleven then, a year older than Brenda, because he got sick when he was twelve. After that I don't remember him ever riding the swings again.

Another favorite activity was the dime toss, in which we threw dimes at goldfish bowls, hoping a dime would land in one and the fish inside would be ours, as well as the small glass fishbowl it was swimming in. These fish usually lived for about a week once removed from their carnival sideshow natural habitat.

We also spent our money on pony rides, sitting on ornery little Shetlands held tightly by Leonard Zehm, who was ready to give a resounding pop of his knee to any miscreant pony as he led us round and round the loop in the park. We all had our own horses, but that didn't matter. It was our civic duty to support the pony rides. We played hide-and-seek in the park among the brand-new John Deere farm equipment parked there on display. Those combine tires were so huge that I could stand hidden, plastered up behind one, without even bending over. When all our energy was spent, we begged fried-chicken money from a parent or a grandparent and sat down with our food to watch a couple of innings of baseball or a bit of the talent show.

It was about six when Mom and Dad would start rounding us up. Their responsibilities were over when Dad had the sound system put away and Mom had the Cake Walk tables folded up. Grandma and Grandpa had already driven back to Spokane. We were tired and filthy. The prizes of the day were in the car, whatever each girl had won, found, bought, or taken as a remembrance.

There was usually at least one cake, some Mardi Gras beads, stuffed animals, piles of chocolate Kisses and coins, and maybe a doomed goldfish gloomily swimming around in a plastic bag of water resting in the glass bowl. In the pickup were the remnants of our parade costumes, that scratchy burlap headed for the burn barrel as soon as we arrived home.

As we drove the ten miles back to the farm that Flag Day of 1960, Marsha gingerly balanced her bagged goldfish in her hands while I cradled a cake on my lap. Cheryl held on to Tracy while Mom drove. Dad followed us in the pickup, which just that morning had carried our child-queen float. He had driven slowly then, about twenty miles an hour, so that the wagon, covered with chicken wire and decorations, would remain upright and those pink tissues stuffed in each chicken-wire hole wouldn't blow out. There was no need for such care on the way home. Liberated from those responsibilities, he drove a flat fifty, with little pieces of pink escaping from the pickup bed to float out across the wheat fields, as he sped toward the farm.

Like Milo Gorton, Bus Sperline was also a teacher. His first job was as a teacher and coach at the Fairfield School, back before they consolidated the districts. Dad and Bus both played baseball and were members of the Service Club and the Park Committee. Bus's wife, Louise, and my mother were friends. The Sperlines didn't live in Fairfield all that long, but those were the baseball years, which seem to have connected my parents to many of their lifelong friends in much the same way that the college years do for other people.

Bus had aspirations higher than those possible in the tiny town of Fairfield. When a job opened up for an athletic director at the new high school in Richland, he nabbed it. In spite of his youth and lack of experience, he didn't have a great deal of competition for the post. Nonscientific types didn't much want to live in the Tri-Cities area, unless they had a contract with the Hanford Engineering Works for a goodly amount of money or were of an entrepreneurial spirit, catering to the whims of the scientists and their colleagues.

Between the scientists and the entrepreneurs, Richland High School exploded in size, an ironic choice of words when you consider the bomb that propelled its existence. In fact, in honor of the town's somewhat auspicious beginnings, they declared that the home high-school team be called the Bombers and emblazoned the name across every young athlete's uniform. They painted a giant atomic mushroom cloud on the side of the school, a huge, impressive mural that reached two stories tall. I think it is still there today.

By the early sixties, Hanford had grown to eight reactors perched on the edge of the Columbia River, sucking in water to cool off the plutonium created inside each giant concrete cylinder. The cooled plutonium was stored and ready to be popped into whatever else went into an atomic bomb, should those Russians push us to the ultimatum we had been threatening. The nuclear arms race was on, and Hanford was at the head of the pack.

We went to visit the Sperlines twice a year, usually staying over at least one night, if not two. We children enjoyed going because Paula, their oldest daughter, who was a couple of years older than Marsha, was, in our child eyes, beautiful. Equally important, she thought we were adorable. Paula had poufed-up hair like Annette Funicello and a soft, melodic voice that could draw us to her like a magnet. From the moment we arrived at the Sperlines' house, Paula cared for us, played Candyland tirelessly, made us lemonade, and told us spooky stories in her bedroom with the lights turned out. Paula was nice, in the way that the word nice was meant to be. She wore plaid skirts and starched white blouses that she ironed herself, got good grades in school, and ushered us children out of the living room when our giggling got too loud for the adults. She walked with grace and laughed lightly at everyone's jokes. My highest ambition was to grow up to be just a little bit as graceful as Paula, not having the courage to imagine I could equal her.

Also helpful to our pilgrimages south was Craig, Paula's brother, who was the same age as Marsha. Craig had a dark-brown crew cut, a stocky build, and, like Paula, a smile that revealed the most perfect set of pearly teeth our side of the Cascade mountains. Unfortunately, as my sisters and I aged, we grew less and less cute in Paula's eyes. At the same time, Craig grew more and more cute in ours, with the unfortunate truth that Craig did not replace Paula's waning adoration of us with affection. He was totally indifferent. Much to our consternation, our gangly bodies and silly-girl personalities were of no interest to him. We knew what we looked like, but also knew that it was just a matter of time before we, too, "filled out," as our parents called it. Certainly Craig should have seen the vast potential that existed in front of him. "Pick a Hein girl! Any Hein girl!" we wanted to yell at him.

Except for the presence of Paula and Craig, we would have detested making that hundred-mile trip to Richland. The drive was excruciatingly long through the dry, empty land that began south of our Palouse and stretched past the Columbia River. I can only remember how hot it always was, peeling ourselves out of the car after the endless drive, and rushing to the oasis of the Sperline house with its built-in air conditioners and iced lemonades. It wasn't just that the town was hot, there was a community lifelessness that stretched out for all the daylight hours. The trees, irrigated by the river, didn't provide nearly enough shade. The sidewalks and streets blistered hot from midmorning on. The view from the Sperlines' house, which sat up on a slope, was of a dry, tan-colored town shimmering in the summer heat. Every house came equipped with an army of air conditioners propped in available windows, filling each home with the small engine roar necessary to provide sufferable air to the inhabitants. It was rare to find anyone on the streets, so tethered to their cooled homes were they. At dusk, though, at about the time the automatic sprinkler systems came on, people would emerge from their houses or return from the Hanford plant and stand on their sidewalks to chat.

I can't think of a better place to build a bomb factory. It was just so hot and weird. Nobody ever talked about it when we went to visit, even though the plutonium processors were just a short drive down the way and almost everyone the Sperlines knew took that drive on workdays. There was still a huge amount of secrecy surrounding the project. With the Russians rabid to ruin the democratic society that God meant for all mankind, the locals couldn't take a chance on foiling things with an idle word or two. The Tri-Cities were created in the shadow of secrecy and easily existed maintaining the silence.

One year we visited at the same time as President Kennedy. We could hardly believe the transformation that had taken place in the town. Red, white, and blue banners hung everywhere, with giant "Welcome, Mr. President" signs in every shop window. There was even a parade in his honor, although the president couldn't attend because of his high-level duties. The townspeople knew he was spiritually in their midst, though, and that was good enough for them. I remember the parade as a lot like Fairfield's Flag Day parade, only much bigger. I was amazed that in spite of the heat there were huge numbers of people. I had no idea so many people lived there or that so many people could be having so much fun in that infernally hot town.

President Kennedy was there for the groundbreaking of Hanford's N-reactor, designed to produce plutonium and then electricity from its excess steam. By the early sixties the Cold War was changing drastically. The Europeans, especially the French, weren't so happy with how the United States was heading up the democracy side, while the Chinese were making waves in the Communist Club. It was no longer an easy case of us against them, and our country's immersion in the Vietnam War only served to exacerbate the situation. Producing lots of raw material for unrealized bombs didn't have the same appeal as it once had. The government decided to convert the reactors to do something a bit more practical, such as making electricity. It was a laudable idea and well worth a presidential visit.

I saw the president on television that evening at the Sperlines'. There were shots of the parade, too—although, unfortunately, none of us on the sidelines—and then there were shots out at the plant, where the president stood on a platform and talked about how important all the work at Hanford had been and would continue to be for our nation. A reactor loomed in the background of the ceremony. It was the first time that I had ever seen one of these peculiar structures, in spite of the numerous times our family had been to Richland.

"What is that?" I asked incredulously. As our nation's president pumped out his public-relations speech, this concrete cone was pumping out huge billows of smoke. Bus looked surprised, as if everyone should know what a nuclear reactor looks like.

My ignorance precipitated a field trip the next day. The president had already flown home, in spite of my wish to get a firsthand peek at Jackie and him, but there were still many people standing in the visitors' area, gazing through the fence at the reactor. I guess the parade made many people want to pay homage to this wondrous thing that existed down the road.

These were pre–Star Wars days, but if George Lucas ever wants a backdrop for yet another sequel, he might consider that structure, minus the fourteen-foot-tall electrified fence with the "Absolutely No Entrance" signs posted every five feet.

There were actually several reactors, but from the road, which was as close as anyone could get without a badge, you could see only one. Most people know what a reactor looks like, but until you stand close to one, and especially if you are a child, you won't understand what a strong impression such a building can make, sort of like death and life at once, embodied in a towering concrete giant. This huge, windowless cement gourd supported a spider web of little blinking lights that zapped irregularly across its smooth surface. I stood there in awe, my eight- or nine-year-old self amazed at the power we could muster against those Communists. From the top, big gray puffs of nuclear smoke sent Morse Code radioactive warnings to those damn Russians on the other side of the world. "You don't have a chance," I read in the smoke that day, standing by the fence. It didn't matter to me that the smoke really wasn't radioactive and that they had already converted the plant to make electricity. Those Russians were getting the message.

This was the early sixties, a good twenty years before that first story broke in *The Spokesman-Review,* Spokane's main newspaper, about the concerns of people living downwind of the reactors. It was even longer before the managers at Hanford started making public, after a great deal of pressure, the secret files that maybe yes, maybe no, helped explain some of our neighborhood problems. In the early sixties, we were just happy to have those giant smoke-stacks puffing away down there.

Eventually the government shut down some of the reactors and converted others to making electrical power. The Cold War waned, and an official Visitors Center was built so that people didn't need to butt up against the fence. Plaques were hung to commemorate the accomplishments of this sweltering trio of towns—Richland, Pasco, and Kennewick—which not that many years before were just part of a larger collection of towns, including White Bluffs and Hanford, towns that many people living in the area today have never heard of.

In August 1963, David Brewer started having terrible headaches that attacked at irregular intervals both day and night. The doctor told Harriett on the phone that the headaches were probably related to allergies or maybe a persistent flu. When they continued she made an office appointment. It was in the middle of harvest, so it was Harriett who took David to the several doctor appointments it took to discover why a twelve-year-old boy would be subject to such misery. Finally, Dr. Thiel did a blood test. This was in the days when the doctor looked at the blood culture himself, as Harriett and her suffering son were in the waiting room leafing through old *Highlights for Children* and Seventh Day Adventist magazines. I think in our times, after looking at that blood culture, Dr. Thiel would have asked Harriett to call her husband, Ed, in for a consultation. The doctor would have told her he needed to speak with both of them, and only when they had each other to lean on, would he have dropped the bomb of a son's cancer. Dr. Thiel, however, following the protocol of the day, ushered Harriett and David back into his office and said, "David has leukemia. Take him to the hospital."

Harriett didn't realize at the time that leukemia was a deadly disease, even more than it is now. Ed realized it, though, when

Harriett went out to tell him in the field where he was cutting peas. Harriett had made a quick lunch for the harvesters, planning to drive David up to the hospital after delivering it out to the fields and speaking with Ed. It was only when she told him about David's diagnosis that she realized her only son was in big trouble.

Ed looked at her, looked at the sky, looked at her, walked away ten yards from the assembled lunch group, and dropped to his knees. With his back to the startled crew, he prayed. Ed was a religious man, but never considered overly so, and this was definitely his first public display of independent praying.

After a few minutes, he pulled himself up, told Warren Henke, his lead hired man, to take over, and said to Harriett, "Let's go." Harriett had driven the pickup to the field. David was already waiting down at the house in the car, curled up in the back seat, his head cradled between his hands.

By that afternoon David had a bed in the hospital and was receiving cobalt treatments. Pat, Harriett's oldest daughter, came home from Pullman, where she lived with her new husband. She took over the harvest cooking and general running of the household. Harriett spent every day in the hospital with David, leaving only in the evenings when Ed arrived after a day in the fields. Ed would sit with David until late at night, dozing in the chair, until he drove the forty-five miles home to sleep for the next day's harvest.

David had a room of his own in the hospital, a place he lay every day in his weakened state. In those days there were no extra beds or lounge chairs for family members, nor were there televisions or the ubiquitous Nintendos found today in pediatric hospital rooms. Harriett passed the time watching her sleeping son, walking the halls, and waiting for visitors to come. People brought gifts—puzzles, teddy bears, and baseball cards—anything they could imagine might make things a little better.

Our Hangman Creek community rallied with delivered casseroles and prayers. Playing at the Dennies' one day, we girls made construction-paper get-well cards; the idea came from Gail Dennie, who had just had a birthday and been given some art supplies. Later that day Dona Hahner, her kids in tow, had stopped by to visit Claudia Dennie, her sister-in-law. Greg Hahner added a card, in which he wrote in his seven-year-old script, "I'll save you a seat on the bus." This got the adults all teary-eyed.

Not us children, though, who really didn't understand exactly why there was so much seriousness revolving around the situation. Mom and Dad were clearly very worried about David, as well as Ed and Harriett. It was hard for Dad to offer much help right then, since he himself was in the middle of our own harvest, a desperately busy time for everyone. We did invite Mary and Brenda over often, though, and Mom drove them to their music lessons. We were all extra nice to the girls, letting them have the first hamburgers off the barbecue and letting them choose their TV trays before anyone else. We had a set of TV trays that had a different outdoor scene printed on each one. I was partial to the tray with the babbling brook rushing down the mountainside and knew for a fact that Brenda was also. Under normal circumstances I set up my tray even before I dished up, announcing, "Dibs," so everyone could hear. But these were serious times, when my natural selfishness was to be put aside. I would still set up the tray, but not say, "Dibs," giving Brenda the opportunity to claim it, which she usually did. It helped me feel that I was doing my part.

After six weeks David was discharged from the hospital. His headaches were gone, as well as his hair, a truly shocking sight for all of us on the bus. He was kind of puffy and pale, but other than that, we just knew he was fine and all heaved a collective sigh of relief.

14 Because my father had no sons he treated his daughters like boys. The pleasure in that came with nightly softball games, a basketball hoop installed at regulation height, and the expectation that we would choose horse 4-H over Campfire Girls. My mother must have exerted her feminine influence to get us to ballet classes in the Waverly Grange Hall basement. There we learned to stand very straight and walk slowly so that our toes touched the floor before our heels did. We each had a tutu of a different pastel color (Marsha, peach; Cheryl, spring green; me, yellow; and Tracy, baby blue). I vaguely remember that we discontinued our ballet lessons when we outgrew the tutus.

We were expected early on to do our share of the chores, not just indoor stuff such as emptying wastebaskets and making beds, but also the same outdoor work any farm son would be expected to do. Our first outdoor work was collecting eggs.

The chicken coop was a small, dank, low-ceilinged building with bad light and the heavy, dry stench of chickens, rotten eggs, and the poop-saturated straw that covered the floor. Against the back wall of the coop was a collection of enclosed square boxes, each filled with straw for a nest. Dad had built one on top of another, ten boxes by four boxes, each with a door that was held shut by a simple wooden slat. Each box, just wide enough for a chicken to scoot in and hunker down, opened in the back to a precarious walkway that led down to the floor.

I hated chickens. I hated them for the way they smelled, the way they walked, the noise the roosters made, and the way they left a thin layer of white, pasty poop everywhere they went. I didn't, however, hate them as much as Marsha did, who, in addition to hating them, was also terrified by them. This fear did not, in my parents' eyes, constitute an excuse from collecting eggs, though, so off they would send us to collect the eggs each evening before supper.

I don't know what Marsha did before I was old enough to protect her from the chickens. Maybe she convinced the hired men to do her gathering for her. I do remember well the days we used to go together. She was ten years old and I was six. Every evening at chicken time she insisted that I accompany her inside the pen. I wasn't as terrified of the chickens as Marsha, but neither did I relish the idea of fending off marauding leghorns while my older sister cowered behind me.

First, we had to feed the chickens, and they knew it. As we edged our way into the pen, the birds would race toward us. To get to the feed room we had to wade through a sea of hungry chickens to the coop, yank the feed room door back hard enough to part the sea, leap inside, and pull the door quickly behind us before any chickens followed. For a brief moment we were safe inside. Then Marsha, older, wiser, and tall enough to reach down into the wooden chicken-mash box, filled the pail with feed and handed it to me. She reasoned that if she had filled the pail, I should go out and distribute it.

From the feed room Marsha would open the door a crack, push the pail and me out, and quickly close the door, remaining safely inside. I was now out in the coop alone with thirty hungry chickens ready to wage war for a taste of what was in my bucket. To properly feed a group of chickens, one should stand in the center of them and gently sprinkle the seed around in a rotating motion, spreading it as far as possible to divide the chickens around the pen. With the chickens madly pecking at my pail and the arms that held it, I pushed my way outside, yelling and kicking at those hated animals. Once outside I heaved the pail of feed in whatever direction it would go, invariably causing a massive chicken pileup wherever the feed landed. I could only hope that the bottom chicken layer didn't suffocate as the rest of the flock piled on, although it would serve them right, I reasoned, for being the greediest of the greedy. My other hope was that my dad never caught me in this rather abbreviated version of feeding time.

Next we had to collect the eggs. Our objective now was to get the eggs and escape from the pen before the chickens finished eating outside. As soon as I returned from leading the flock outside, Marsha emerged from the feed room and joined me. She could have begun collecting the eggs while I was out feeding, except for one thing: there might still be a hen sitting on a nest and Marsha would have to face it alone.

Chickens like to do their egg-laying in private, a respectable wish from one of the world's dumbest creatures. Apparently, once they get started, even the lure of food can't drag them away from an egg half-laid. Maybe laying eggs kills the appetite. Whatever the reason, we could never be sure upon opening a chicken box that there wouldn't be a big fat hen hunkered down over her eggs. One by one we opened each box, praying to find them empty except for an egg or two. On occasion, though, we would crack a door open and find the light from the back opening blocked out by a giant hen. We would slam the door shut, flick the slat closed, and rush through the rest of gathering, praying all the time that the hen would not emerge from the back before we could get out. I don't think we had a clear idea of what would happen if a chicken caught us alone in the open room, but we didn't want to find out.

Collecting eggs never became easy for us, which meant it never became easy for our parents, who had to nag and cajole us to get out there every evening. Poor Mom, her nagging couldn't even draw our eyes away from Tarzan and Jane, immersed in the treetop coy talk captured on our RCA black-and-white. "At a commercial," we would repeatedly throw over our shoulders until Dad arrived. Dad was a barker. His staccato "Get! Out! There! Now!" could pop us off the couch right in the heart of an African quicksand scene, sending us to our own personal quicksand scene—the chicken coop.

Our attitudes, not to mention the ever-present stench of chickens that emanated around our farm on breezy days, was enough eventually to persuade our parents to do away with those pathetic birds. The adults must have been planning it for weeks, if not longer, but I only remember the day itself. Grandpa and Grandma Hein came down from Spokane, and Grandpa and Grandma Keller came out from Fairfield. Russell even came from wherever he always came from to work on the assembly line, or disassembly line, as the case was. We children were there, too, of course, although we were no help at all, perched around the edges of the event, watching in utter amazement as Dad entered the chicken pen, grabbed his nearest victim, which squawked and tussled as Dad threw it on a wooden block. In one fell swoop Dad axed off the chicken's head, letting it fall limp to the ground, while the body quivered and rolled spasmodically.

Never again would the expression "running around like a chicken with its head cut off" have an abstract meaning for me. Chickens really do run around with their heads cut off. Cheryl and I, who witnessed the first slaughter up close, squawked ourselves, in utter horror, as we ran away, certain those shaking, headless bodies were chasing us. Did we really think the chicken, or what was left of it, could see us? Did it think we children, who desperately hated to feed them, were in some way responsible for this morbid fate to which their entire flock was condemned? We weren't taking any chances, parking ourselves on top of the nearest wooden fence. From there we sat and watched as Dad and Grandpa Hein took turns grabbing and chopping, one by one, until that entire flock was nothing more than a pile of heads on the right and the quivering bodies on the left, or wherever they had given up the ghost, probably not all that far from the chopping block. Cheryl's and my collective memory still has them all at the foot of that fence, right beneath us, in one last attempt at flock revenge, certain as we were that they held us accountable for this genocide.

Once all the chickens were dead, Cheryl and I climbed down from the fence and followed a wheelbarrow full of dead birds to the shed, where the grandmas and Mom were already plucking a previous load. Wet feathers were everywhere, both the big old stiff ones and the soft ones from underneath. The chicken bodies were dipped in giant pots of boiling water to make the feathers come out more easily. Bald chicken bodies, with their rubbery pink-gray pimple-marked skin, lay end to end in a long row stretching down the workbench. Grandpa Keller and Russell gutted them, slicing the stomachs open and pulling out handfuls of gooey body parts that got divided into one of two categories: gooey edible parts such as gizzards and hearts, which got set aside, and the rest, which got slopped into buckets. The most amazing moments were when they found partially formed eggs inside, the shells soft and pliant and barely able to contain the makings of baby chicks. After they gutted and cleaned the chickens, they put the bodies in plastic bags and sealed them up to be taken to the locker we rented in town. There was no way the whole flock would fit in the freezer in the basement. A dull, heavy, wet stench of rot and death hung in the air for days.

Shortly after that, Dad shoveled and hosed out the chicken pen, hooked up a watering trough, moved out the boxes, replaced them with a manger, bolted to the wall fertilizer barrels that our saddles could sit on, and moved the horses into the former chicken coop. The straw on the floor took on a horse-manure stench, and in no time, the feed room smelled like fly spray, oats, and old leather, a perfume to any self-respecting farmgirl.

Between this new horse barn and the Quonset shed was a drooping wooden building we called "the old shack." Inside were sawhorses, used pipe fittings, coffee cans filled with rusted bolts and nails, defunct life jackets, a wobbly picnic table, a number of giant metal cans that smelled of solvent, ten or twelve iron fence posts (the kind with the barbs on them for wrapping wire), and a pervasive odor of must, mildew, and mysterious chemicals. There were also huge, complicated spider webs strung thickly across just about everything, with an occasional golf-ball-size spider—motionless, hairy, and horrifying to my sisters and me. It was because of the spiders that we were not particularly sad when Dad announced that he was going to tear down the old shack. He wanted to replace it with a trailer where hired men could sleep during harvest.

Russell had been our favorite hired man forever, and if tearing down the old shack meant Russell would be moving in with us on the farm, we were all for it. Hired men were so familiar I always thought it peculiar that they didn't all live with us. During the summer they were on the farm from dawn to way after dark, ate all meals with us, and even took showers in the basement. Only on occasion did they sleep over, and that was always in the heart of harvest.

I was in the house watching *My Friend Flicka* when Dad, while tearing out the shack's rafters, found the birds. Cheryl raced in, out of breath, to announce that Dad had uncovered a nest of baby sparrows in the eaves. Her delivery of this news could be compared to announcing the bombing of Nagasaki. It was big. It was now. And, as I was soon to find out, it was to be fatal for the birds. Cheryl had a premonition of doom in the way she yelled the news throughout the kitchen and living room. "Get out there now!" she screamed at me, when I was waiting for a commercial to start.

Cheryl was usually the picture of calm. At eleven years old, she was tall, skinny as a rail, and flat as a pancake. She had a bit of a dancer's body, when I think of it, and sort of floated through life. She never went hysterical on us, so her screaming foretold something big. I left the television even though the show was still on.

I remember the desperate screeching coming from across the yard, a heart-pumping, terrorized plea bursting out of every vocal cord of those little birds. There they were, I discovered after climbing the ladder, a nest of five of the scrawniest, baldest, ugliest living creatures I had ever seen. They looked like giant, dirty, pink insects. Most striking were their mouths, or, more properly put, their beaks, which stretched tightly across their entire faces and remained open into caverns that obscured the rest of their heads. I was certain the beak size was directly responsible for the outrageous noises bursting from their heads. With bodies that could fit neatly inside my child hand, those chicks could put out a thunderous chorus of birdie racket.

"Don't touch them!" Marsha screamed as she raced across the yard, fresh from the horse pasture, where she had been grooming Shauna's hooves. Mary Brewer followed closely at Marsha's heels, having ridden her bike over to help Marsha braid Shauna's mane. Those two knew from their eighth-grade science class that one scourging flick of a human finger on one baby bird and the mother would have nothing to do with the baby birds ever again. As Marsha explained this to us, Mary backed her up with serious nods.

I was not so concerned about the chicks. I had inherited our family bird phobia. Mom was traumatized in her childhood by having to sleep in an attic where, according to her, bats swooped down on her every night. She told the story over and over to her children, adding more bats and numbers of swoops with each telling, until we were left to believe that Mom had spent the majority of her childhood nights not in slumber but, rather, quivering in terror under a thin blanket in the attic of some North Dakota parsonage. Combining that story with the way Mom would cower and run whenever a bird flew by us instilled an abnormally high distaste for birds in me, her child most susceptible to dramatic stories. I hate the idea of a bird sitting on my shoulder, distrust anyone who chooses a bird for a pet, and even, as already mentioned, am fairly uncomfortable about collecting chicken eggs.

Marsha and Mary's ornithological contribution about mother abandonment to the present crisis only made me more disgusted with birds as a species. If my little finger dropped for one second on the featherless body of one of these screeching beings, its mother would fly off into the horizon. On the other hand, I couldn't blame the mother of a nest full of things so grotesquely ugly for wanting to fly away. However, responsibilities are responsibilities, and if she was going to hatch these babies into the world, she had to take care of them, no matter what. I had the idea that this "finger on feather" stuff was just an excuse the mother bird kingdom used as a way to move on to greener pastures, or bluer skies, as the case may be.

Dad was not very interested in the ornithological world according to Marsha or Mary. He was much more interested in tearing down the old shack. So he reached up into the nest and pulled the little birds down one by one, causing Marsha and Mary to add their voices to the screeches. "It doesn't matter," Mom yelled above the din. She had arrived to comfort us. "Even human beings looking at the nest has caused the mother bird to abandon the babies." Marsha and Mary fell silent. Their teacher hadn't presented this piece of information at school. Even *looking* at the nest? The degree of disgust I had for birds escalated.

Now, in my greater understanding as an adult, I'm not sure if we would call this a white or a bald-faced lie on Mom's part. For the record, we do not generally think of our mother as a liar, and I forgive her in this case because she wanted to protect us. Not me so much, since by this time the expression "That's for the birds" had taken on new meaning for me. I didn't really care what happened. But my older sisters—Ms. Audubon and Ms. Nagasaki—did, not to mention our neighbor Mary. Mom knew the survival of the little birds was pretty hopeless, whether Dad killed them or Marsha tried her hand at feeding them. And it wasn't even that the girls felt a particular affinity for birds. They were just at a stage when they were especially outraged at anything unfair to the animal world.

"He's going to kill the baby birds?" Cheryl looked incredulous.

"How?" I wanted to know.

"Ahhhhkkkkkk!" Marsha said, unable to articulate the extreme emotions she had for the bird kingdom at this moment.

"Their mother has abandoned them," Mom pointed out.

"We could feed them worms," Cheryl suggested.

Marsha then regained her voice enough to explain about mother birds chewing things up and then regurgitating them for the babies.

"What does "regurgitate" mean?" I asked.

"It means," Mary piped in authoritatively, "they puke the food back up into the babies' mouths."

At that point I was finished with the topic forever. I knew my place was with Flicka and the black-and-white TV in the living room. Horses were a species for which I had a great love and respect. I could count on their loyalty, and their eating habits were similar to my own. So when Dad took those baby birds and threw them like miniature baseballs against the shed wall, ending their short lives quickly like little snuffed-out candles, I didn't hear a thing. At that very moment the TV music that always plays loudly at dramatic moments had come on, and Flicka was galloping at breakneck speed across the countryside to rescue young Ken from the rattlesnake that had him trapped up against a rock.

Because Dad and Mom thought ownership would foster a sense of responsibility, they gave each of us our own horse, which we were to feed, currycomb, clean its hooves, and exercise. Responsible behavior does not come naturally to young people, and the ticket was to find horses that we genuinely would grow to love.

In the early days, there were a few false starts with Shetland ponies that we chose for their size, a fact that belied their appropriateness as children's pets. In fact, these cute, pint-size horses possess poison personalities generally bent on the destruction of little farmgirls. We were bucked off, bitten, and kicked countless times. It didn't mean we didn't protest each time Dad threatened to sell the ponies, but it did mean that most of the time it was Dad, not us, who threw hay into their mangers or combed the burrs out of their manes. When he sensed their heads reaching around to take a bite of his thigh as he labored over them with the currycomb, he could swing around and knee them in the jaw, grabbing onto the pony's mane for support. With his gimp leg and eroded sense of balance, Dad was unsteadied easily. The bite that reached his thigh was level with my shoulder, and a pop of my fist only caused the pony to rear back and scare me more. So Dad fed the ponies and eventually sold the ponies. Only thanks to Mom's insistence did we get a second chance with other breeds.

I personally learned about responsibility from Rockette, a mostly Welsh mare we bought in Sandpoint, Idaho, from an ad in the Spokane newspaper. She cost three hundred dollars, which included her saddle and bridle. Rockette was a little plump, her rounded belly attesting to her age and numerous pregnancies, and she had a spectacular personality. By spectacular I mean she didn't bite, kick, or buck and tried hard to be everything I wanted her to be.

Sometimes I wanted her to be a thoroughbred, especially during that period when I was reading *Man O' War*, the first hardback book I ever owned, a true story about a racehorse. Every morning in the summer I would put Rockette's western saddle on her, while pretending it was an English. Cheryl on her horse, Lady Anne, and I on Rockette had races to the top of the first hill on the dirt road. Lady was a taller horse than Rockette, and I would start a bit ahead. Rockette ran her hardest, morning after morning, although we rarely won.

Sometimes I wanted her to be a rodeo pony, and she dutifully circled the gymkhana poles that we'd spaced out in the back pasture. I kicked, yanked at the reins, and whooped accordingly as Rockette trotted back and forth. She wasn't very good at it and consequently knocked most of the poles down, but she tried hard and I knew it.

Sometimes I wanted her to be a circus pony. This was probably the hardest for her, as I practiced standing on her back, balancing on one leg, and rolling under her stomach as if she were a balance beam. She stood there, solid as a rock, resisting any temptation to flick a fly with her hoof, as my ten-year-old head hung precariously underneath her belly.

At other times I let Rockette be the Indian pony she was, and I became Whiet-alks, Qualchan's beautiful wife, loping over the fields to tell my people that the whites had killed our chief, my husband. It was a sad fantasy. Usually the people needed me to take over the tribe, and I had to rise above my own personal grief to keep our people alive.

Our membership in the Cayuse Kings and Queens encouraged my sisters' and my passion for horses and any event featuring them. The only requirement to be in the club was to have a horse and to have the club uniform: a cowboy shirt, a green bolo tie with a horse's head for a clasp, Wrangler jeans, cowboy boots, and a straw cowboy hat with a green ribbon the same color as the bolo tie. We needed to have matching outfits for the Flag Day parade, which we actually rode in one year.

The parade was the highlight of the club's existence—our way to display our passion for horses to the whole community. We practiced for weeks for that parade appearance, riding up and down the gravel road in pairs in front of our leader, Dick Dennie's, house. A couple of mothers even made a cloth banner with "Cayuse Kings and Queens" stitched across the center in giant green letters. Lynn Dennie and Sally Felgenhauer carried it stretched out between them from atop their horses. Those girls had tall, spirited animals—Lynn, an Appaloosa, and Sally, a quarter horse–Arabian mix—horses that were even a bit hard to handle, which reflected well on our whole club. The rest of us, mounted on our extremely easy-to-handle horses, lined up two by two behind them, paired by Dick for reasons we were not aware of and not about to question. He was a little cranky that day. Nerves always ran high on Flag Day, at least before the parade.

In our imaginations our troop pranced regally down Main Street, hooves playing syncopated rhythms on the pavement, as the crowds marveled at our formation and spirited beauty. The truth was that most of our horses weren't shod, so they made more of a dull thud on the pavement than a crisp *clip-clop*. Our formation, whatever existed of it, lasted about one block before each horse took to plodding along at its leisure. My friend Mary Rohwer and I were at the very end, I think now because our horses were the smallest and the slowest. We started out the parade sitting erect and proud in our cowgirl finest, but as Rockette and Thunder tuckered out in that June heat, Mary and I lagged farther and farther behind the rest of the group. It was all we could do just to keep the horses from stopping completely, as we kicked, hissed, and prodded them along. We were eventually passed by the Worley Badgerettes, a girls' marching team from across the Idaho border, and we, along with Rockette and Thunder, became a couple of gals on horses plodding up the street, rather than integral members of the Cayuse Kings and Queens.

In good weather I would usually go out into the pasture in the evening to wish Rockette a good night. It was my favorite time of day. In the waning light the alfalfa took on a deep, rich, dark-green color. The air was full of the sound of crickets, which chirruped from mysterious places we never found. Off in the distance I could hear the drone of Jimmy Hahner's tractor, or a car going over the highway on the horizon, or a crop duster getting in one last turn. Then there was the smell, of course, which changed according to the season. There might be the smell of new alfalfa, fresh-cut hay, wheat ready for harvest, or cotton-wood-tree buds, sappy, ready to bloom, and oozing an odor that claimed the entire farm.

One evening when I was eleven, I arrived to say good night and found Rockette to be indifferent to me. Scratching her ears barely drew a response. Her head sunk low to the ground and her eyes were dull. When a carrot went unaccepted, I sneaked a fistful of oats from the cow barn. When she refused that too, I knew something was very wrong. Mom said we would check on her in the morning.

The horse pasture was immediately outside my bedroom window, and Rockette, lying in the field between the lilac bushes and me, was the first thing I saw the next morning when I awoke. I dressed quickly with a foreboding knot in my stomach. I think I said good morning to anyone who was in the kitchen as I passed through on my way to the pasture.

I didn't even get through the gate before I knew she was dead. She was lying on her right side, her legs extended as if she had laid down to roll over on her back in that way she did now and again to get the flies off or scratch herself. I couldn't bear to go near her and just stood there hanging on the gate, wondering what I could possibly do next. Shocked and certainly feeling quite empty, I finally turned around, walked myself into the house, and sat down for breakfast.

Over breakfast I learned that Dad had to take a cow and one of the dogs to the vet in Cheney. He needed someone to go with him and I was elected. The cow pasture was on the opposite end of the farm from the horse pasture, so Dad and I were able to go outside, load up the cow, lure the dog into the cab, and drive away without so much as glancing behind the house toward the horses. We drove all the way to Cheney, had the cow tested for pregnancy, got the dog shot up, and filled in the vet on farm gossip without me so much as whispering one word about Rockette. But I remember my heart. The words *broken, aching,* and *heavy,* which had often prefaced the word *heart* in books, were all appropriate for me at that moment, and I knew it. My stomach hurt, my whole body hurt, and my heart was broken.

Regardless, I would not say a word, because if I did, what my heart and my stomach already knew would be really and irrevocably true. There would be no more circuses, no more rodeos, and no more thoroughbred horse racing because Rockette was dead.

Two hours after we'd left, Dad and I drove back to the barn. As we backed the truck up to unload the cow, I heard the screen door slam. The circus was over. There was Mom walking toward us from the house, crying. She said through her tears, "I have very bad news, Teri." I snapped alert like a soldier at "atten-shun!"

"I already know!" I screamed, racing across the farm to the horse corral, where Mejroc was. He was Rockette's oldest son, a grown gelding and ornery as sin. I buried my head in his neck, sobbing. Rockette was now, and forever would be, dead. As I sobbed, my tears and running nose clotting against the dirt and loose hair of Mejroc's shoulder, he reached around and nipped my upper arm.

It's usually no big deal when an animal dies on a farm. They die and, if we didn't eat them, we'd call a truck to take them away. Rockette's death was different. My whole family heralded her as the circus queen who could withstand any physical discomfort in honor of the make-believe world of farmgirls. For this reason my father did not call the truck, but instead called the man in Fairfield who had the backhoe. Dad paid him twenty-five dollars to come dig a grave for Rockette in the horse pasture outside my bedroom window. I spent the day in my bedroom listening to the sound of that backhoe, grinding and honing away a place for Rockette to rest permanently. Periodically my mother came upstairs to see how I was doing. Usually she would bring me something—a tuna sandwich or a cookie—and for that day only, this saddest day I'd yet known in my life, Mom allowed me to drink a whole can of Coke instead of sharing one with Cheryl.

In the mid-1800s, in the interest of settling the West, the government granted huge tracts of land to the railroad companies. Public land was sectioned off in 640-acre pieces and numbered. The railroads were given the odd-numbered sections to encourage them to build railroad lines from the eastern states to the western territories. The even-numbered sections were kept by the government, many quartered and then offered as homestead pieces to those willing and able to carve a home from them. Northern Pacific got the largest railroad land grant: 39,000,000 acres stretching from Lake Superior to Puget Sound. They received twenty odd-numbered sections for each mile of right-of-way across states and forty odd-numbered sections for each mile of right-of-way across territories.

The railroad could sell any sections they didn't need, which they did in abundance in the Palouse for about $2.60 an acre. Homesteads were cheaper, although a bit more complicated to obtain. Heads of households, men or women over twenty-one years of age, could apply for a homestead, paying out about $22 in fees for the paperwork. After applying, they needed to live and work on the piece continuously for five years before a homesteader could apply to "perfect" the homestead. Only at this moment, when the clerk recorded this in the Tract Registry, did the homesteader actually own the land. My Great-grandpa Hein was signed off on November 8, 1879, and Great-grandpa Thams, who homesteaded three miles north of the Hein Homestead, was signed off on April 21, 1890.

In our neighborhood, the Thams homesteaded, the Smiths had railroad land, the Kegleys homesteaded, and the Kenos bought railroad. We have small homesteads out here because the soil is so good. In places like eastern Montana the homesteads stretched out to 320 acres, with the hope that those farmers could make a go of it on that desolate land that was so often not arable. Most couldn't, even on twice that size. We were lucky we came West when we did, when there was still Palouse land left, if I can count myself among the "we" of my family.

Even though my grandparents grew up only three miles apart, this was a fair distance, and they went to different schools. Grandpa went to one that still sort of stands down by the Waverly-Plaza Road. It has three upright wooden walls with the fourth caved in and a roof that lunges south, making the whole building appear lopsided, as if it were caught in a fun-house mirror. Still clinging to the top of that tippy roof is the casing for the school bell, although the bell itself is long gone. Nobody can seem to remember the name of that school for sure, but Elsa Keno thought it might have been Curlew, a name to honor the tens of thousands of birds that used to inhabit the area.

Grandma went to Rattlers Run School on the north bluff above Hangman Creek. For a decade or more, Rattlers Run School was headquarters for most of the organized activity that occurred around Hangman Creek. Not only was it the school and the locale of Spelling Bee Fridays, but it was where the "literaries" met. A literary was a cross between a modern-day book-club meeting and a debate. People, both men and women, came to the school once a month, at which time a question was posed to the group; for example: What led to more sorrow, alcohol or war? What is worse, a cranky, neat wife or a happy, messy one? Should there be compulsory military training in schools? The evening was spent debating the chosen topic, and no one was allowed to leave until the group had arrived at a consensus.

The school was also where the gypsies came. Their arrival was greeted with ambivalent feelings by the pioneers, who believed the gypsies to be thieves but, nonetheless, very entertaining performers. Among them were magicians, storytellers, and ventriloquists who arrived the night before a performance to practice and sleep in the school building. Word traveled fast when the gypsies were back, and everyone came for the show. The homesteaders paid them with loaves of white bread, gutted prairie chickens, and dried fruit from the year before.

My great-uncle learned to juggle thanks to a gypsy whose name he never knew. Hans Jr. was probably the one in the family who most looked forward to the gypsies' arrival and begged his mother one year to sew him some juggling balls. When the gypsies next came, Hans was ready. Most of the gypsies didn't speak German or English, so he approached one and started to throw the sacks in the air, letting them dramatically plop to the ground to illustrate the point that he couldn't juggle at all. The man ignored him, but from across the meadow an old woman came. Taking Hans's three sacks in her dirty hands, she looped them up in the air, one over the other, up and around and over, up and around and over, and finished the demonstration by making each sack disappear behind her back. She smiled a toothless grin and walked away. My great-uncle was young—Grandma couldn't remember exactly how old—but she did remember that he stood there and cried at his introduction to highway robbery.

The old woman turned just when she reached the other side of the meadow. Hans saw that she was tossing one ball in one hand, up and down, into the same hand. Then, from the other hand, she added a second ball, up and down, just the two balls in and out of the same hand in a monotonous routine. Finally, she added the third ball and the second hand, and she was juggling. For one minute she did this, and then she sent each sack flying his way and walked off before he could pick them up. He practiced for months and actually became quite good, eager for her return the next spring, but he never saw her again.

In the early 1920s, they moved the school's location to Big Flat, a stretch of land on the road to Waverly where the Fourth of July celebration had been held for years. I'm sure my family attended, because the land adjacent to Big Flat was part of their piece.

From the June 30, 1882, edition of the *Northwest Tribune, The Pioneer Farmers Paper of Eastern Washington:*

> The citizens of Hangman Creek, and surrounding countryside, will celebrate the Fourth at the grove on Big Flat, two miles south of Smythe's Ford, on Hangman Creek. Ample preparations will be made by the committee of arrangements for the accommodation of all. Good speakers have been invited, and a large platform will be erected for dancing, swings, base ball, and other amusements. The committee will spare no pains to make it one of the most pleasant gatherings that has ever come off in Spokane county.

On her mantle Grandma Hein displayed a 1926 school picture of my aunt, Burnice, and my dad at Big Flat School. The teacher in the photo was Erma Grant, according to the chalkboard sign held up by two of the students. There are five boys and five girls in the photo. In 1926 my father was six years old. Three of the five boys in the photo are wearing overalls instead of the jeans the other two are wearing. Each boy holds at his side his wool cap, except the boy at the end, the one next to my dad, who had dropped it behind him. The girls are all wearing dresses. Burnice is holding the sign along with a stubby kid in a checked shirt. She looks a little smug at her huge responsibility. I presume the girls wore their best cotton dresses, the kind with what we now call Peter Pan collars, and leather boots laced up to cover legs wearing thick, black stockings.

My father can't tell me much about Erma Grant, not even how long she was his teacher. He just can't remember, claiming always that all the trouble in his life, and all the medication that has warded off bigger troubles for all of us, has left him with a memory resembling Swiss cheese, irregular, with unpredictable gaps. My father's memory is vague, but he first went to school at Big Flat and then to Liberty School, which sits where the present school is. After Liberty, he went to Waverly School and then ended his school career graduating with a class of twenty-two from Fairfield High School. He never moved once; it was just a process of settling in as families decided how many schools to have where.

We considered our Hangman Creek neighborhood a rough two or three square miles, an oblong-shaped section that stretched out away from our house more to the northeast than any other direction. The miles include rolling hills, mostly fields, with the farms of our neighbors resting in the corners, the alcoves, and the pockets between the fields. Within this area are ten farmhouses. Of those, some are farmed by the descendants of original homestead families, whereas others are farmed by family friends or people with connections in some way to the original owners. Farmland never goes for want of a farmer, and one can often trace a lineage of some sort between a farmer and his fields.

Old people lived on two farms. Ervin and Esther Zehm lived where my great-grandparents had homesteaded, moving there to make room on their piece down the road for their son, Leonard. Uncle Detlef and Aunt Mary lived just over the hill from us.

Leonard looked after his parents, my parents looked after Detlef and Mary, and we children looked after one another, riding our horses to secret meeting places to investigate mysterious rodent holes, or over to someone's farm to play "kick the can" or "red rover." On special days we would play "shock the baby," a game in which we neighborhood kids would line up holding hands in a long, skinny line snaking out through somebody's pasture. We were in no particular order except that the youngest was on the end. On the count of three, the person closest to the electric fence would grab onto it. The current would ripple through us, growing stronger and stronger as it passed through each human-conductor kiddy-body. We jerked, screeched, and giggled uncontrollably for a few seconds until kids starting letting go, rippled away from each other in this electric madness. There is no direct evidence that anyone suffered irreparable damage from that game, although our parents did discourage it. We saw it as a wild way of being brave.

We children all rode the school bus together—a tiny vehicle with five double seats on each side, driven by Lola Hagen or Joe Clifford, our bus drivers for years. During the school year we made our game plans on the school bus, and in the summer we made our plans over the telephone. Of course, we all were on the same party line.

The party line had far-reaching utilitarian purposes. One winter in the early fifties, when the snow was particularly bad and people had been snowed in for days on end, my dad told everyone that Dick and he would ski to Fairfield for groceries. At ten in the morning, my mother, who was pregnant with Cheryl at the time, sat at the phone and took orders. She held the phone to her ear as one after another neighbor waited their turn to put in an order. "Baking powder, sugar, a package of steaks from the freezer locker, okay," she would say. Or "Tapioca, Quaker Oats, salt, okay." Or "Ten pounds of flour!" Mom said, writing down "five." It was ten miles to town and another ten miles back, as Dad and Dick skied and pulled the sled to each party-liner's house. It was a very long day, and by the end the two were party-line heroes.

For years we considered the party line a public forum, open to anyone who wanted to listen in. No one knows when phone etiquette changed. At some point it was no longer acceptable behavior to eavesdrop on other people's conversations. Our parents batted the term "rubbernecking" around with some frequency, laying down the law whenever we got caught hunched over the phone with our hand over the mouthpiece.

My sisters and I developed a technique for lifting the receiver in absolute silence, gently placing our hand over the mouthpiece so the speakers wouldn't hear our breathing. If Jimmy was calling Cornwalls for parts, or Harriett dialing her mother to chat, we could just as silently replace the receiver as if we were never there. We always stayed on the line if any of our friends were calling another friend or if there was any indication of good gossip.

We especially liked to pick up the phone when Barbara Martinson was on. The Martinsons lived over by the Brewers, right where our Kegley Quarter is bordered on the east by Kentuck Trails Road and on the south by Hays Road. Barbara and her husband, Joel, really knew how to have fun. They were younger than our parents and hung around with a wilder crowd in Fairfield, at least until Barbara got sick. They occasionally drank until they got tipsy and often stayed out past midnight, both qualities that were somewhat racy to us.

The Martinson kids—Belinda, Diane, and Stevie—were younger than all of us, and Barbara often hired me to come over and baby-sit when she and Joel went out. I loved baby-sitting for them, not because the kids were great—they weren't—but because I got to dictate which television shows we watched, and because the Martinson house had things lying around that I would never find at my own house. There were stacks of weathered Playboy magazines under the television, 45s lying around of rock-and-roll music I'd never heard before, and that little plastic toilet on the kitchen table. It was about five inches tall and sat between the salt and pepper shakers. Around the bottom of the base was written "Good-bye, Cruel World." The lid opened up, and inside was this little plastic man, halfway flushed down the toilet. He had a satisfied look on his face, like the whole thing was intentional. That toilet always cracked me up, sitting right there where everybody ate their breakfasts.

The Martinson kids were hellions, although I quite adored them for their spirit. They raced around the house screaming, biting, and generally causing anyone who cared more than me a great deal of misery. My tactic was to ignore and indulge them, a plan that worked fairly well, with a surprising lack of breakage. When they got really out of hand, I just locked them outside until bedtime. Getting them to bed was not as difficult as you would think. As soon as I yelled, "Bedtime!" the three raced for their room. They shared a large bedroom in the back of the house where three twin beds were lined up down the north wall. All three knew that once they were in their beds I'd go to the kitchen and return with a tray on which would be giant bowls of ice cream topped with chocolate sauce and marshmallows. Bedtime was their favorite part of the day. At thirteen I didn't know that the reason the three of them got so weird and jumpy after the ice cream was probably a sugar high, but I did know that once they were done they would get all crazy again. I edged toward the door as they finished up the ice cream, waved a pleasant good night, and put a chair against the door behind me when I left. An hour or two later I would return to the bedroom, pick the sleeping children up one by one from the floor where each had landed, and tuck them into bed. Then I cleaned up, shelved everything they'd thrown around the room, threw out the ripped coloring-book pages, wiped any sticky ice-cream drips as best I could, turned off the light, and left the door open a crack behind me, all to give Barbara and Joel the impression that I was carefully monitoring their children. The kids were under strict rules from their parents not to have dessert unless they behaved themselves, which, to my knowledge, they had never done. I think Barbara and Joel must have wondered at the enormous amount of ice cream that I, a relatively small teenager, consumed. I was glad they stayed out so late, long after I'd nodded off on the couch in front of Johnny Carson, since I not only earned more money, but it also allowed me the several hours necessary to get the kids to bed properly.

When my employers finally did get home, Barbara would come in, usually laughing about something, hand me some money, and say, "Joel is in the truck." I always felt kind of nervous around Joel, even though he was perfectly nice to me. I think it was his goatee and the dragon tattoo that curled up on his right arm—blue, red, and mean—that made me nervous. You never know what somebody with a tattoo like that is going to do next, or so I thought at the time. Usually he just talked on and on about something that had happened that day, as if I were an adult and we were regular friends. I never said anything, but that didn't seem to bother him.

Joel kept a shapely little plastic doll in a scant bikini dangling from his pickup-truck mirror. The doll winked one eye and wiggled whenever he hit a bump on the drive to my house. The doll didn't look anything like Barbara, although she didn't seem to mind. Nothing seemed to bother her. Both she and Joel had great senses of humor and were always cracking up at one thing or another.

Barbara never gave me the details about their nights out, but by listening to her and her friends on the phone, we neighborhood children learned conclusively about the swinging life of a downtown Fairfield we had no other access to. We overheard Barbara announce that Marlene Hensley, who was the RN at the nursing home, had gotten so tipsy at the Steak House one night that she passed out in her car as it warmed up out front. Barbara said that Rick Horner got slapped by Patricia Linstrom on the dance floor, and we could sure guess why, knowing what Rick could be like after a couple of drinks. Another time she told Rhoda Hengen how Norm, who owned the Steak House, had these business cards made up that read "Norm's Big Meat Is Waiting For You." Norm's wife, Francine, got really mad and made him burn the whole box of them, although Gail and I, who were listening in on that one, couldn't figure out what the big deal was.

When Barbara got lupus I kept going over to their house, but then it was to help clean house and iron. Her kids were too young to help much, except to change the TV channels for her. I didn't have to baby-sit them much in those days because Joel and Barbara mostly stayed home. I didn't understand what lupus was, but I observed that it seemed to make a person really tired and kind of grouchy.

We kids lost an eavesdropping gold mine after Barbara got sick, although we still had Pamela Reedy. I don't know how the Reedys ended up on our party line, because they lived almost in Waverly. If we found Pamela on the phone, we figured we'd hit the jackpot. Considering the present-day popularity of 900 telephone numbers, Pamela was ahead of her time. She came as close to phone sex as a nice Lutheran girl in the early sixties could, especially when you consider that, in her late teens, she was most likely still a virgin. That didn't matter. In her sultry, innocent voice she lamented to the boy of the moment that she wasn't right there next to him to give him a big kiss. She said she would do just about anything to show him how much she loved him. She uttered "anything" in a sort of breathy little-girl voice that every kid in the neighborhood could imitate. As far as we know, that's as far as her phone sex ever got—kissing and "just about anything," but those sentences were as titillating as anything we could imagine, which, frankly, at the time, wasn't much. "That Pamela sure has got a lot of love inside of her," I would say to Cheryl with a wink.

"Have you been rubbernecking again?" Mom would demand.

"No," I said, disgusted at Mom for even thinking to accuse me.

Sometimes we breathed too heavily, or accidentally chuckled, and whoever was talking would say, "Now you kids get off the line!" Maybe we would and maybe we wouldn't. The two parties speaking with each other would spend the rest of their time punctuating their conversation with "Do you hear breathing?" or "Is someone on the line?" They always asked this in a really loud voice, as if the rubberneckers were deaf or something. Luckily there were enough kids on the line that blame couldn't be pinned on any specific one. Besides, essentially, we all were guilty.

When the doctors diagnosed David Brewer with leukemia and he was in the hospital, his sisters rode the school bus home to our house, often staying until late, if not overnight. The Brewers' phone rang into our house also—our ring was two quicks, while theirs was one long and a quick. Mary was appropriately worried and strained, and when the Brewer phone rang, we marveled at how fast she could bolt down the stairs from Marsha's bedroom to our living room where our phone sat on the desk next to the television.

My mother told me years later that she lived in dread of the phone ringing on those endless nights when David was in the hospital. She focused her concern on the possibility that the news would come over the phone when David's sisters were at our house, finding that easier to worry about than something so incomprehensibly horrible as the possible death of our neighbor boy.

During David's last week, that March 1964, callers didn't linger, wanting to keep the lines open for Ed and Harriett to call their girls. As a ten-year-old, I didn't quite understand. When David initially got out of the hospital and looked so bald and pudgy, Mom had said that he would get back to looking normal. I took that to mean he would be okay. Well, he never did either— look normal again or get okay. This was before the "tell all" philosophy of child rearing, so we were spared the details.

My strongest memory of David is of a fat, bald, and pale boy. Now when I think about it, underneath all that illness, he really wasn't any of those things. He really was a daredevil on the Flag Day swings, a nice kid to little Greg Hahner on the school bus, and, as Harriett told me years later, a huge baseball fan. She still has his catcher's mitt. I also remember not liking him after he got sick, although I probably didn't dislike him so much as felt uncomfortable in his presence. My parents had told me a thousand times that David was very sick and that I needed to be extra nice to him. I was nervous even sitting by him on the school bus for fear that his proximity would force me into an unnatural kindness, or maybe I was afraid I wouldn't be nice enough to him.

That feeling of discomfort grew after the time I accidentally bopped him on the head with my spelling book as I flung it across the bus toward Gail Dennie. Mrs. Tread was the substitute teacher that week, and she gave full-size Hershey bars to the top two winners of the Friday Spelling Bee. Gail and I were in training that morning as the bus lumbered toward school and the contest. You would have thought I'd dropped an anvil on David's head for the bus driver Lola's reaction. The bus came to a screeching halt, right there in the middle of Kentuck Trails Road, and I received a royal chewing out from Lola. I thought at the time that David and she should both have been glad it was the half-inch spelling book I threw. My math or English book could have knocked him out cold.

I kept that opinion to myself, however, until I was in my mid-twenties and ran into Lola at church one Sunday when I was visiting home. So many years after David's funeral Lola couldn't remember the incident. She apologized nonetheless for overreacting, although pointed out that I was far from an angel on the bus. I guess David made everybody nervous in those days.

David died when I was in fifth grade. I know that because I remember going out into the hall for a drink, around the corner from the fifth-grade classroom, and there, in the middle of the morning, were Harriett and Ed, walking in front of me toward the principal's office. They were all dressed up and holding onto each other in an unusual way. Even though Ed often gave me piggyback rides, pulled my ponytail, and told endless kid jokes, I stopped in the hall and watched them in silence. You almost never saw somebody's father at school in the middle of the morning, especially in a suit. Maybe in overalls, but never in a suit.

I had moved over to the fountain to get my drink when Mrs. Fox came out of the classroom and called to Ed and Harriett. I will never forget the two women collapsing into each other's arms, nor the way they cried right there in the hall during school hours, their shoulders kind of hunched over into the sobs. I had never seen anything like that and immediately figured out that David was dead. Ed just stood there stiff and somber in that suit while the women wept. He looked my way, but I don't think he saw me. Then Brenda came out of her classroom at the other end of the hall. As she trudged alone down the hall toward her parents, I thought she walked as if she were going to prison.

My mom wanted her daughters to go to the funeral. She said, "You'll never believe that he's dead unless you go." But we didn't want to, and I guess she didn't feel like forcing us. I just couldn't imagine what a funeral would be like or how I could bear to see David Brewer dead in a coffin. Now I am a bit sorry I didn't go and very sorry that every time I conjure up his face, it is bald and pudgy, which is how he was for only six months of his whole life. I do have a few other memories: the Flag Day swing image—

David, a blur going round and round through the Fairfield air; a memory of the back of his taller self in the front seat of the school bus next to the smaller back of Greg Hahner; a vague recollection that he teased me before he got sick; a memory that we had more than one game together of "kick the can," and, at least once, we held hands for "shock the baby."

* * * * *

When I think of it now, David was my first cancer kid. He was the first of hundreds over these years, and I have become their defender and protector, much like Lola was David's protector that day thirty-five years ago on the bus. Whenever I'm feeling angry at the odd looks healthy kids give my cancer students now, I try to remember how awkward David Brewer made me feel, and how I must have looked at him sometimes. Now I'm so accustomed to bald kids that I don't even remember if they have hair or not when I meet them—the same way it's easy to forget if some men have beards or not after they're out of the room.

My students are definitely peculiar to look at: all swollen from steroids, completely bald, with skin that is a pallid pinkish yellow, often crisscrossed with railroad tracklike stretch marks. Their hands shake from their medication, which makes holding a pencil difficult for them. Many insist on writing for themselves, however, commanding that piece of leaded wood to make letters the way it is supposed to for a ten-year-old student.

In some ways my young charges are like palsied old people—tired, listless, and shaky all over. I have to remind myself that somewhere in there they are still children, and then I think that I should remind them also, since they must forget sometimes. I have loaded children in the most pathetic conditions into my car for field trips, choosing places where there aren't too many people who might stare at my students' masked faces and bulbous bodies, or the portable IV pumps that fit over the backs of their wheelchairs.

I promised eleven-year-old Brian from Montana a trip to the Boeing Flight Museum with his doctor's approval, if Brian promised not to swear at the docents, an urge that it turned out he could, in fact, control. Being so sick had made him very cranky, and he held nothing back from anyone, except on that special day. The elderly docents treated us like royalty, swinging the doors open in a majestic way, while I pushed Brian through in his chair. Brian loved looking at the airplanes suspended from the ceiling of that giant museum and decided on the spot that he would be a pilot. I can't help but think the image of flying away must have been overwhelmingly seductive to him.

Before the doctors decided face masks didn't protect anyone from germs, Amy from Alabama and I were treated by the management of Seattle's seventy-six-story Columbia Tower to a lunch in the exclusive club located at the very top of that building. Amy bought a new dress for the occasion, as well as a scarf for her head, and insisted on riding the elevator to the top without the help of a wheelchair, even though she was tapering off steroids and her joints were killing her. She wore a mask over her mouth, gracefully pulling it down every time she took a bite. The club members pretended we were invisible, although our waiter treated Amy like a movie star and me like her agent. Amy told me she would remember that meal for her whole life, which, if I remember correctly, wasn't that much longer.

Only once, over these years, have I had a field-trip problem, and that was when Jason, the large boy from Samoa, passed out during an IMAX movie at the Pacific Science Center and those wonderful teachers from Prosser helped me prop him up until the 911 people came. He was okay, just dehydrated, but the paramedics insisted on giving him IV fluids and returning him to the cancer center clinic in the ambulance, instead of in my car. I remember how sad I felt as Jason, now fully conscious, was wheeled off. He was seventeen and comfortable with the idea of riding in an ambulance with strangers, but we both felt bad. Our cover was blown. Right there in the IMAX theater, getting ready to watch a forty-minute movie about Alaska, he had passed out, right in front of those Prosser schoolkids, many of whom had already looked at him oddly out of the corner of their eyes before the lights went out.

Maybe my small mission in life is to create normal moments for sick kids and to help others consider that underneath all the physical manifestations of illness in a youngster is just a kid wanting to watch a forty-minute IMAX film like every other kid in the theater. And when I'm really feeling sentimental, I like to think that David Brewer, from whatever heavenly place he is in, notices that Teri Hein, the girl who crowned him on the school bus with that spelling book, has finally learned her lesson.

On a farm there are many things to drive: the pickup, the combine, the swather, the wheel tractor, the Cat tractor, the trucks, and, of course, the cars and the riding lawn mower. Farmboys learn to drive at age twelve. Their father farmers initiate them on the wheel tractor that they drive snail-like through the hay fields, pulling the trailer full of bales. Considering the circumstances, it wasn't at all surprising when Marsha turned twelve that Dad put her on the metal seat and taught her the three things she needed to know about driving: how to turn the steering wheel, how to make the tractor go faster, and how to stop it.

Driving a tractor that is pulling a trailer is not easy. It is even harder to maneuver this dinosaur through a skinny-ribbon draw carved between two wheat fields that is obstacle-coursed with hundred-pound bales.

My father is not a patient man. He never has been. Ever since they took out his thyroid gland he has had to take lots of medication to keep his body tides ebbing and flowing. Sometimes my mother thinks medication buildup makes him so cranky. That might be true. Or maybe he's just cranky sometimes.

Marsha had a difficult time avoiding bales in the field. Daily practice did not seem to pay off, and with each trip out to the draw you could count on Marsha to roll a trailer wheel on top of at least one bale, if not two, or three. When this happened, the doomed bale would pop open, the twine stretched past its limit. My father, stretched past his limit, would yell. Marsha, accustomed as she may have been to his impatience, would weep, jump off the tractor, and race dramatically for home, her cowboy boots slipping on the dry straw left loose on the swathed fields.

You could consider hers an overreaction since Dad never stayed mad long. We all had seen *National Velvet* several times, however, and were partial to Liz Taylor's dramatic exits when things weren't going right.

My sister Cheryl's experiences as a twelve-year-old tractor driver didn't go much more smoothly. By then Marsha had been retired to the kitchen to help Mom make pies and pot roasts for the hired men. By nature Cheryl was calmer than Marsha but not a bit better at driving. The results were very similar. From my vantage point at the kitchen table I could see her walking up the draw in that somewhat jerky way people walk when they've been running and they're too tired to continue but still too upset to move slowly. "Uh oh," I would say to Marsha and Mom, "looks like we have another broken bale."

I don't think I realized at the time that I should have been cherishing my youth. I am only eighteen months younger than Cheryl and loved to spend those hot June afternoons in the cool of the TV room nursing a Coke. Sure enough, two haying seasons later, we all were sitting at the kitchen table eating Frosted Flakes when Dad announced he was taking me out to the tractor.

Good-bye *Donna Reed Show* and Paul Peterson, you cute thing, you. Good-bye, Johnny Weismuller *Tarzan* reruns. Good-bye, piecrust-dough leftovers baked with cinnamon and sugar into little circle treats by Marsha and Mom, and left on the counter for any child hanging around the kitchen. Good-bye, lazy afternoons riding on the back of the trailer as the tractor pulled it bumping through the draws.

If the weather was cool enough, I loved to ride out to the hay fields on the trailer, climbing up the pile of bales as the hired men stacked them higher and higher until everything, the bales and me, headed for home, up and over the hills, wobbling precariously with each rut in the field. Sitting there, so regal in the very middle, I hung onto the twine of the bales on either side as if they would steady me. Maybe I would be holding a kitten that had come along at the last minute, both of us looking out over that view from the height of ten bales. There it was, the Palouse, the draws, curving through the fields, dotted with bales destined for pickup on the next run. Off on the eastern horizon were the Tekoa Mountains in Idaho and, to the south, Steptoe Butte.

That time of year the wheat fields were almost a lime green, the heads threatening every day to turn yellow. The pea fields stretched for acres, with plants holding skinny, green pods not yet ready for me to pick. When they were ready, my mom always gave me the top of the double boiler to fill with fresh peas. We ate our fill, dinner after dinner, coated with Blue Bonnet margarine, not ever making a dent in the fields that surrounded the house. Sometimes I would let Tracy come along with me. She could usually be counted on to pick about ten peas before beginning to whine to go home. Then the season was over and the remaining peas dried on the vines. The combine later scooped them up to become that stuff in plastic bags on aisle five at Safeway that people use to make split-pea soup. But that would be weeks away. During haying time the pods were flat with only small bumps.

When Dad announced my promotion to the tractor seat I felt a bit like a new pea pod. I was not quite ready to be plucked off my trailer, nor, as the case was, away from my bowl of Frosted

Flakes. Good-bye to the view and the great pleasure of haying season. With several thousand bales stretching out every which way, just daring me to break them, all the attention of this twelve-year-old girl would be given to the task at hand.

Kids learn on the wheel tractor because there is only one pedal—the brake pedal. The throttle is on the column, and the steering wheel is fairly manageable. At least that is the perception farmer fathers have of the vehicle. To a twelve-year-old girl who much preferred to be in the house drinking Coke, currycombing her horse, or watching *Queen for a Day*, a wheel tractor is covered in grease and dust, has brakes that grab if you look at them, possesses a throttle only Hercules could move, and features a steering wheel the size of a wagon wheel and more difficult to turn.

The metal seat alone could dominate a whole paragraph. The factory had custom-shaped the big dish, we all assumed, to fit the giant butt of John Deere. It was shiny and slippery with an odd piece of metal humping up in the middle to fit against your crotch. The seat didn't have a thread of padding. I guess John expected you to bring your own. It was hot, extremely hot, sizzling, in fact, to a farmgirl in cut-off jeans as she delicately placed her tiny rear in the middle of the giant dish, trying in vain to keep the sides of the hump from burning the inside of her thighs, which, of course, happened every time she moved or hit a bump or turned around to see what Dad was yelling about. The whole experience was horrible. It was dusty. There were thousands of bees. The air was still, hot, and shimmery. And Dad had given a Zen lesson in tractor driving: just do it.

I couldn't sit in the seat and push the brake at the same time. In fact, I couldn't sit in the seat and do anything else at the same time. I had to perch precariously on the hump in front of the seat, grip the steering wheel with one hand, steady myself with the other on the seat behind me, and practice pumping the brake up and down to get the tractor to stop.

My twelve-year-old foot shook on the brake pedal as Dad hooked on the trailer. I slowly let it out a fraction of an inch and the tractor lurched forward at five miles an hour. For me it might as well have been seventy-five miles an hour. I pumped the brake in a frenzy. I was terrified. Squinting through the heat, I heard nothing but my heartbeat. All of a sudden I noticed that I was crying. Dad hadn't even yelled at me yet, and I was already crying.

"Oh, Jiminee," he said, exasperated and disappointed. "Go get your sister." I crawled off the tractor in defeat, sniffled my way into the house, looked up at Marsha, and said, "Dad wants you outside."

She looked at me in pure disgust. Even though I recognized my utter uselessness as a farm daughter and knew my older sisters had every right to despise me, my will to live was stronger. I smiled as I flipped on the television, just in time for the end of the *Donna Reed Show*. Don Drysdale, the Dodger pitcher, was making a guest appearance.

Resurrected from the kitchen, Marsha took her place on the wheel tractor seat, her sixteen-year-old legs gracefully reaching to the brake pedal below. Later that very day, as he was about to throw another bale onto the trailer, Dwight, one of our hired men, tripped in a hole and fell in front of the trailer's back tire. Marsha, still perched on the tractor seat, the steering wheel in her confident grip, responded to his yell by slamming on the brake, bringing the trailer to rest on top of Dwight's upper thigh. As he screamed in surprise and pain, Marsha's foot froze on the brake, unable to lift it in spite of Dwight's noise. It took Dad leaping up behind her and knocking her foot aside to move the tractor forward and roll it off Dwight's injured leg.

A teenage neighbor boy's legs, we learned, hold up well under trailer tires. They are much sturdier than, say, farmer father patience levels. While Dwight returned to haying in only three days, my father's daughters never did, except to resume our regal places on top of the bales, which is where we really belonged, anyway.

I never knew my father without a limp. After his first bout of headaches following his surgery, the limp was just there, slight but constant, enough for him to give up dancing, golf, and baseball, but not bad enough to give up farming, at least initially. His left leg was slightly slower than the right, and, if he got very tired, it was way slower, causing him to drag his leg, as if he had just risen from a cramped position and the leg was completely asleep. In fact, since his thyroid surgery, that leg was always in some state, either partial or complete, of being asleep. As with the headaches, the doctors were at a loss to explain the possible connection between the limp and his absent thyroid gland, but there was no doubt that the two were somehow connected. When the gland went, the limp came, accompanied by the headaches. Unlike the headaches, the drag of his left leg never went away. Because of my post-operation birth, I never saw my father play league baseball, go hunting, cross-country ski, dance the jitterbug, or run the relay race at Flag Day.

In 1962, ten years after the surgery, Dad and Mom drove to Seattle for a physical. It seems that the surgeon in Spokane had attempted an innovative approach to the curing of thyroid cancer by cutting out bits of just about everything my father's neck contained. The big leagues of cancer, over on the other side of the mountains, wanted to have a firsthand look at this surgical work of art, more familiarly known to us as my father's neck.

Dad hadn't really had any particular troubles since the surgery, except for the rocket headaches that were now long gone, and, of course, his gimp left leg. He took then, and continues to take now, an arsenal of pills that assist his body in doing whatever that long-gone thyroid used to do.

His balance was affected, which affected his golf and his bowling. The microscopic way it got worse each year slowly eroded the pleasure he took in these sports, until finally he gave them up completely. After the surgery and the long recovery time, he never went back to skiing or baseball. My mother said she didn't mind that the drooping leg broke his rhythm when dancing; he could make up the difference every couple of measures. Besides, it brought his dancing ability closer to her level, about which she had always been so self-conscious. It mattered to him, however. Dancing went the way of golfing and bowling as each subsequent year proved that his leg was not going to recover and, in fact, worsened.

These were small prices to pay for the eradication of his cancer—the gimpy gait, the crabbiness, the diminished enthusiasm, the arsenal of drugs he popped daily, usually at breakfast, and the jagged scar on his neck, which, unlike his limp, faded a little bit more each year. "Lucky to be alive," was Dad's chorus, and Mom's, too, as they drove back from Seattle with the good news that, ten years after the surgery, there was no sign of cancer and he could, for all practical purposes, consider himself cured.

This was one year before David Brewer got sick and a few years before Mona Zehm's brain betrayed her. Her symptoms were definitely, if nothing else, peculiar. Leonard noticed one evening while they were eating that every time Mona raised her fork toward her mouth, she had an odd way of hesitating for a few seconds before she would continue to eat—holding that forkful of food in midair, as if not sure what to do with it. He asked her about it, but she didn't know what on earth he was talking about.

The kids noticed it too, but her insistence that she didn't know about it seemed to imply some kind of mini–time warp for her. After observing this eating behavior for several meals in a row, Leonard became genuinely concerned. He called Mona's parents, who lived down by Waverly. "I'd love to come over and watch Mona eat," joked her mother, Gladys, who found Leonard's concern more unusual than Mona's behavior.

That evening Gladys, along with Mona's father, Claris, insisted on making a huge dinner. Sure enough, Mona's odd hesitancy occurred, as had become usual. It was just odd enough to be eerie, and both parents, along with Leonard, decided Mona was due a trip to the doctor.

The next day Dr. Thiel suggested an inflammation of the brain could be the root of the problem, and nothing to be greatly concerned about. To be on the safe side, though, he recommended they see an ear, nose, and throat specialist in Spokane. Because Mona felt perfectly fine, an appointment three days later was acceptable. That evening her "perfectly fine" turned into a nightmare when she had her first seizure, a wild thrashing attack that occurred right after supper. To contain her, Leonard had to hold her with such force that her shoulder was dislocated. It was the only way he could cease her violent movements. He held her there on the floor, as the family stood watching in horror. It was Gladys, who had stopped by for another evening, who noticed Mona's face turning blue, and who realized that she had swallowed her tongue. Grabbing a soup spoon from the dinner table, she pried open Mona's mouth and pulled her daughter's tongue out from her throat, saving her from asphyxiation.

When Mona came to, she was no more aware of the seizure, or the swallowing of her tongue, than she had been aware of her dinnertime hesitations. She was forty-one years old at the time. Her oldest son, Dwight, was off at college by then. Her youngest child, Polly, was nine years old and had just finished third grade.

Mona's brain tumor was so large they could take only about half of it without turning her, as Dr. Thiel put it, "into a complete vegetable." Dr. Thiel wasn't the surgeon, but he had been at the surgery and acted as spokesperson for his patient and friend, Mona. The surgical team had told Leonard that she could be perfectly fine after the surgery, or she could die, or she could be left in any imaginable state in between those two possibilities.

As it was, she was left with her left arm completely paralyzed and her left leg partially so. Only thanks to the rigorous exercise that Leonard insisted on, he himself pushing that leg back and forth for hours each day, did Mona retain some feeling in it. Her speech was a bit slurred, although it wasn't too hard to understand her, and she had a gigantic scar that cut across her head where her hair would never again grow quite right. That surgery transformed Mona from a vibrant, gregarious, athletic person to a broken woman.

Mona and my mother were about the same age and were very good friends. The Zehms' farm was close to ours, but we were each on the edge of a different elementary school district. The Zehm kids went to grade school in Spangle and we went to Fairfield, the school district lines passing down the Cahill Road between our farms. We did, however, see each other every Sunday at church, where Leonard has been a deacon for as long as I can remember.

We kids were pretty well lined up in age. Marsha and Dwight were the same age. He was her first prom date and the one who she ran over with the haying trailer. Then the Zehms had another boy, named Gary, and then Barry, who was the same age as Cheryl. I came along, and then one year later they had Keith, and finally Polly was born the same year as Tracy. Of course, everybody in the neighborhood had four or five kids, and it wasn't so coincidental that we all happened to be about the same age. When you thought about it, most everyone's ancestors had come out west at the same time in the late 1880s, and the ancestors were all about the same age; it takes a young, strong person to make that kind of trek. Then, with families passing down farms from generation to generation, we all ended up in more or less the same graduating classes.

Mona was a dynamic woman with a loud, alto voice and a booming laugh. She was tall with permed, curly, dark-brown hair covering her head. Before her illness I remember her mostly wearing pedal pushers and sleeveless blouses. She came over to our house often to trade stories with my mother, bringing one or two of her boys and Polly in tow. The boys always wanted to ride our ponies because they didn't have any. They were wild kids, pulling our kittens' tails, spooking the horses, and picking my mother's flowers when they weren't racing around in circles on our bicycles. Mona was constantly reining them in. She didn't beat on them or anything, just laid down the law in a way that we all knew meant business. She didn't scare me, though, because she was always joking and sweet to us girls, probably because she had this whole slew of boys, until Polly finally came along. Before that, though, she would usually give my ponytail a yank when she saw me. "Hey, Little Missy," she would say. Sometimes I sat on her lap.

Leonard was softer. He was not much taller than she was, and when he laughed his shoulders went up and down in time to the "ha-ha-has." His parents lived in the house down by the creek that my great-grandparents had built. After Hans and Catherina died, and after Leonard grew old enough to take over his family farm, my family sold our old homestead to Leonard's parents,

Ervin and Esther. Ervin just died this past spring. I think he was pushing a hundred.

When Mona got sick the whole neighborhood, and especially we children, had our ideas of illness redefined forever. Every other sick person we'd known had either recovered or died, but Mona seemed frozen into this in-between state between alive and dead. The doctors said she would never really be well again. The best we could hope for was that Mona would have a halfway life with only a whisper of ability to perform any of the duties, responsibilities, pleasures, and functions of a wife, mother, and person in the world.

While Leonard was in the field, the Women of the Church took shifts that summer of Mona's surgery—cooking, doing the laundry, propping her up, and instructing each of her legs to move forward as they propelled her toward the bathroom. Leonard had a special leg brace made for her that didn't allow her to walk unassisted, but did help her to stand up alone. He also bought her a wheelchair and added a porch onto the house so she could sit in the fresh air and take in what there was to see. The women left as dinner warmed in the oven and Mona sat in her chair to await the evening return of her husband and the hired men from the fields. She watched as her sons set out the food for dinner, and she watched as they washed the plates after. She did the most she could, which wasn't much at all, at first. When Leonard finished his supper he got them all, including Mona, ready for bed. She could push her right foot up to make it easier for him to take off her shoe, and she could cup the toothpaste cap in her hand. With time, Mona learned to maneuver around with her wheelchair, enough to actually do quite a bit of housework. Her children learned to be helpful by choice, and the family pulled together in a unified show of strength.

I don't remember Mona ever coming over to our house again, although I'm sure she did. We went over to their house often after that, but it wasn't really like going to their house, since the new Mona barely resembled the robust Mona who pulled my ponytail. Since her speech was slurred, I had to strain to understand her, and the hand she used to pull my ponytail was all curled up and resting in her lap. I was thirteen by then, and the ponytail was long gone. I couldn't imagine any child ever sitting in that lap again since her hand was always there.

Mona never seemed angry about her condition. She was always so glad to see us, propped up in her wheelchair in the living room. My mom would come in and pour lemonade for all, bending the straw in Mona's glass for easier drinking. The women sat and visited, while we girls fidgeted on the couch and ignored the women urging us to go outside and play with the boys. Tracy and Polly were good friends, Polly being the same age as Tracy. Those two would run off together to look for kittens or play house in the boys' abandoned fort down below the willow trees. But Cheryl and I would just sit frozen on the couch, paralyzed ourselves in our own pubescent way.

If you don't count picking dandelions for my mother (one cent per hundred), my first job was roguing wheat for Jimmy Hahner. The Hahners' house was the only one we could see from our house. There it was, perched up on top of a hill to the northeast. Our fields sat up against each other in many places. Our Kegley Quarter rested against their Marion Piece. Aunt Burnice's forty acres was up against the old Goff homestead. Our Metcalf Place ran into their Wernz Place down by Kentuck Trails Road.

Fields have names, of course, like skyscrapers or legislative bills. Generally speaking, you not only know the names of your fields, but also those of your neighbors. Of course, it isn't as if one farm is one field with one name. One farm family usually farms several fields, each with a name most often based on who home-steaded it or divided the section. Most of the names are so old we can only guess where they came from. My family, for example, farmed the Rohmer Section, the Kegley Quarter, the Lemon Place, Burnice's Piece, the Creek Place, and the Flat.

Field boundaries can be kind of deceptive. To somebody driving by, the landscape might look like one giant field, rolling over and over the hills toward the Tekoa Mountains of Idaho.

In fact, the lines are drawn by a telephone pole, a draw, or maybe a county road marking the end of the Kegley Quarter (nobody remembers who they were) and the beginning of Louise Perryman's Place. (Louise, who hadn't lived on the property since her teens, died of old age many years ago.)

We couldn't see the Hahner house from ours, although we could see the Hahners' new house once Jimmy built it. Note the distinction between the Hahner house and the Hahners' house. If they were old enough, houses had names. Jimmy's dad had built the old house in the early part of the century. By the time I was born, this lumbering structure was two stories of peeling white paint, a swaybacked roof, and a wraparound porch with steps on two sides. From one set of stairs you could get to a lopsided sidewalk that led out to where cars were parked. The other led to the trail that curved out to their barn and shed. Both these buildings were tucked into a dip between the hills at the end of a long driveway off Kentuck Trails Road.

The plumbing, and most likely everything else, was messed up in that old house, and in one year, although it seemed overnight, Jimmy built Dona a new ranch-style perched on top of one of the hills. The new place had giant picture windows that looked out over the Palouse. People couldn't get over that view when they had the open house. It wasn't as if people hadn't seen a view like it before; half of the people there had something similar out of their own windows. Maybe it was the way the carpenters had framed their view with those picture windows.

The new driveway cut down through a field and passed by the old barn and shed. Jimmy tore down the old house soon after so that it wouldn't fill up with rats, mold, and cold the way Louise Perryman's Place had. Every year her old house had listed more to the east until one day it just eased over onto the ground, kind of like it was sitting down slowly into an easy chair. There were old photographs still in the attic that ended up spread out all over. I thought my dad, who farmed for Louise, had stuffed the salvageable ones into a Manila envelope and sent them off to her son, Roger, who lived in Montana. But years later I found a few in a drawer in the back of our shed. They were yellowed pictures of big Germans in uncomfortable clothing and sourpuss faces who nobody knew anymore.

After Jimmy built the ranch-style and they had the open house, he built a garage of corrugated metal next to the new house. Then he landscaped a square of grass in front of both buildings. He planted a couple of fruit trees and a small flower bed and called the place finished. The house sat on the tallest hill in the area, and for the first time since my grandfather built our house, we could walk out into the front yard and see that other people had homes on the planet.

It didn't bother us, seeing another house after all those years. We liked the Hahners. Their eldest son, Greg, was two years younger than me, with three more siblings after him: Ronda, Bryan, and Jill. We kids rode the same school bus; we all went to the same church; we all had been baptized by either Grandpa or Pastor Mueller, who came to preach after they called Grandpa to Oregon. Dona and Mom rode to Ladies Aide together on Saturdays, and Jimmy and Dad played dice at the Steak House on Friday mornings during the winter. We felt just fine about being able to see their house from ours.

Jimmy and Dad took many a Coke break together whenever they were working in adjacent fields. For hours, their Cat tractors, like a small agricultural drill team, disked round and round adjacent fields, puffing up the soil, working in the manure, getting it ready for planting. Then, as if on cue, the tractors would come to a simultaneous stop. Dad and Jimmy, in the black-and-white overalls all the men wore, the stripes barely visible under the grease and layers of dirt, would jump off. White skin ringed their eyes. Their goggles did a good job of protecting that skin from the grease and dirt that otherwise covered them from top to toe.

Mom or Dona, or both, drove out in pickups, bumping up through the draw that divided the two fields. Each had Cokes, tightly wrapped in white freezer bags, maybe a sandwich, and a cookie or two to accompany this midafternoon break. The men leaned up against the pickup, commenting on this or that, and the women did the same. The women were usually more lively, their minds not addled by all the circles, the noise, and the dust made over and over again by those tractors.

Jimmy gave me my first job, roguing wheat, as well as my sister Cheryl, and Jan and Gail Dennie, who lived on the other side of the Hahners. The word *rogue* has three meanings. It means "tramp" as a noun, "vicious" as an adjective, and, as a verb, "to weed out diseased, inferior, or unwanted things from a field."

The Grain Growers Association paid farmers a lot of money for wheat free of unwanted things. Flawless wheat commanded a high price as seed. To qualify, one's field had to be free of Canadian thistles, button weeds, dog fennel, wild oats, henbit, bachelor buttons, and, especially, the scourge of all wheat fields: morning glory.

To an outsider, morning glory is a lovely plant. Its delicate vine curls up to soft leaves and white flowers that reach up toward the wheat stalk to which it gently clings. To a girl whose job it is to clear a field of the plant, morning glory is an insidious murderer strangling wheat stalks one by one. It wraps itself around the wheat plant, growing tighter and stronger as it grows, twisting itself into a gnarled mass, sticky and prickly— and almost impossible to disentangle for any girl with a short fuse and cumbersome gloves. If only it would have been completely impossible to disentangle. Then Jimmy would never have asked us to do it. We could have just clipped the wheat stalk off with a pocketknife or maybe a pair of kitchen scissors, ridding the field of victim and attacker in one fell swoop. "Oh, no," Jimmy said, "twist off the morning glory and leave the wheat. You can do it," he told us every morning at six-thirty as he dropped us off at the field.

Jimmy was a nice man, the kind of man for whom you wanted to do a good job. He was rather quiet, unassuming, and unintimidating. Of course, we knew all the men in the neighborhood, and none were the least bit intimidating to us, at least in the way we knew them to be at that point, as neighbors and pseudo-uncles. But we also knew these neighbor men could be gruff with their hired men, and we neighbor girls weren't interested in experiencing the gruffness of any men other than our own fathers. We accepted the roguing job because we felt confident that Jimmy wouldn't be cranky with us. Besides, in my case, I had my eye on a lime-green vinyl coat like Goldie Hawn's on the television show *Laugh-In*, and my mom said I'd have to finance something so impractical myself.

Jimmy was a bit of an anomaly in his family. His last name attested to the fact that he came from a long line of area farmers. His cousins, brothers, and parents controlled big parts of the county and were the heads of church councils and service clubs all over the district. The Hahner family, in its extended sense, was as close to a dynasty as our area came. The husbands were board members at the Grain Growers Association. The wives were advisers for the Rainbow Girls and held offices in Ladies Aide. The teenage boys were track stars, and the teenage girls were cheerleaders. Jimmy was many things in Fairfield, on the Grain Growers board, on the Wheat Growers board, as well as head of the church board for some years. He did all of his community service in such an unassuming way, however, that no one would ever have called him power-hungry or bossy but, rather, a good man doing whatever was his civic duty. He probably just wasn't interested in the big-fish swim through the microscopic pond of Fairfield. Even as a child I thought he was too kind to be powerful, although I doubt if I understood the significance of kindness at the time.

I don't remember him ever teasing me, like Ed Brewer did. I don't remember him ever raising his voice, even to his own children. I don't remember him ever complaining about anyone. I just remember that he smiled a lot and was very kind—an easy person to have for my first boss. I knew if we didn't work well, he would probably not tell us. I imagine Jan, Gail, and Cheryl were thinking the same thing when they accepted the job.

This is how we worked—the Hein/Dennie Girl Brigade—lining up abreast on the edge of the field. We waited for the cue from Jan, at thirteen the oldest of us, to move forward. Given it, we pushed slowly through the already tall and golden wheat stalks, which rose to our midthighs, or to mine, at least, as I was the smallest of the four. The bachelor buttons were easy to pull up. The button weeds weren't bad and generally grew in clumps that came up with one yank. The morning glory was definitely the worst. Gail and I prayed that the inevitable clusters of plants would fall into the pathway of Cheryl or Jan. They were older and didn't seem to take it quite so badly. I think they even relished the challenge. As the two forewomen of the crew, they took their roles seriously. Moving optimistically through the fields, they assiduously unwrapped morning glory, grabbed at thistles and button weeds alike, and urged Gail and me on. We also had our roles. We were the younger sisters and the whiners. We took our time, intentionally overlooking clumps we didn't think anyone would notice, breaking off whole wheat stalks with morning glory twisted around, and asking Jan, who had a watch, what time it was about every fifteen minutes. We knew that, at eleven, somebody's mom would come to pick us up. By then the sun would be beginning to scorch, and we would be caked with dust and starving, that bowl of cereal having worn off hours before. It was really awful, although I don't think we ever considered quitting.

Though quitting wasn't an option, whining was. As the morning progressed you could often hear my voice or Gail's wafting across the field, followed by either Jan's "Shut up, Gail," or Cheryl's "Shut up, Teri."

We did this work for two whole weeks. Back and forth we moved through that field, bandannas wrapped around our necks to soak up the sweat. Our cowboy boots were hot and grimy inside from shuffling through the dust, and our once-white anklets had slipped down under our heels, covered with stickers and almost off our feet up into the toes of each boot. Our long-sleeved shirts and long jeans added immeasurably to our discomfort, but protected us from the thistles that grew in patches throughout the field. We had scissors for the thistles, or Jan did. She delicately cut the base while we gingerly pulled up a sticky leaf and heaved the whole plant as far away as we could, back into the section we had already rogued so we wouldn't meet up with it again. It would dry out and fall down dead before the men at the Grain Growers checked the wheat.

The two weeks seemed like a lifetime, but finally we finished and Jimmy gave us cash—twenty-five dollars each. I think the field qualified for seed, although now that I think about it, Jimmy never again asked us to rogue his fields. The school bus went by that field twice every day, and for a year or so I looked at it with a mixture of familiarity and contempt. Even these days I glance over when driving—surprised at how small the area was that we worked—knowing the boundaries because the telephone lines are still not buried. It's called the Roberts Piece. Even though different people farm it now, I'm sure it goes by the same name.

We were probably the most united as a family during winter. Mom did not go to school because the roads were too precarious. The fields were frozen, and Dad's farming duties were limited to maintaining the machinery. His realm was the shed, the barn, and the house. We had an intercom system set up between the shed and the house, the gray box sitting by the refrigerator, its microphone dangling from the side. Static would fill the kitchen to tip us off that Dad was trying to communicate with us. "Hello!" he yelled through the box, his voice in a distant well across the farm. "What time is dinner?" or "Call Reinholt and order a sprocket," or, the worst, "Send out one of the girls." Often that meant Dad was welding.

Dad could weld anything to anything, and the only things he needed were his welder, a machine that sent off sparks like a Fourth of July rocket, and a daughter to hold up some heavy metal part for an endless amount of time while he painstakingly melted it and something else together. Dad had a hood and special gloves that he wore when he welded, standing close to the project and looking like a deputy from hell sent to our shed on a special mission. He required whichever girl arrived at the shed, bundled up in layers of wool to confront the cold, to hold up the metal thing completely unprotected from the sparks that flew every which way. Dad thought because we stood away from the sparks, and technically did not ever get hit by them (except maybe once or twice), we didn't need any special protection. "Just look away from it" was all he said as he lit the torch. It was sort of like being under the drill at the dentist in that tense way we waited for the machine to finish its dastardly work, knowing at any moment a searing hot spark could zap us, or we could be blinded by an inadvertent peek.

The shed was a giant Quonset hut over by the horse barn. In the winter it housed the combine, the wheel tractor, the pickup, and anything else that needed tending. Our basketball hoop was also there, as well as two walls covered with tools, a wood stove to provide a little heat in that corner, and an ever-present cat or two trying to cash in on the warmth the wood stove put out. The shed had giant roll-back doors on both ends that hung open in the summer but were squeaked shut to a permanent seal for the length of winter. A tiny door on the south end let in a whoosh of winter air during December as we who were called into service by the intercom came in and out.

We hated being called to the shed. It was cold outside, which required the ritual of pulling on every piece of warm clothing we owned. Once we headed out to the shed, Lord only knew how long Dad would require us to stay, propping up the metal piece for welding. If, as our muscles began to ache, we let the piece droop a bit, Dad would snap at us to keep it up. We would hold, he would bolt, we would droop, he would yell, we would boost it up, and then he would be done.

Dad could be a real bear out there. It was a sure thing that if you got the call to the shed you would probably be moved to tears at some point by Dad, who didn't seem to understand that preteen girls are not necessarily the best shop partners. It wasn't that he hit us, or even yelled loudly, it was just the quality of his utter contempt for our foibles that was so emotionally devastating. He had this way of hissing through his teeth whenever he was disgusted, a leaky sound that made us feel like the flat tires we were. Girls like to please their dads, and we were really lousy at it out in the shed. In spite of the abundant evidence of this fact, he always called us out. We were all he had, and even when we finished whatever task he wanted us to perform, he kept us on retainer while he fiddled with something else.

During this time Dad let us shoot baskets. We dribbled our way around the Cat tractor to practice our lay-up shots or stood behind the pickup bed to practice free throws. The basketball hoop we got for Christmas years ago still hangs in the shed.

Thanks to winter and welding, we all are fairly good basketball players today, known to outshoot from the free-throw line many a mediocre guy. When Dad was in the mood, he would treat us to a game of "horse" or one-on-one, which made the trip out to the shed worth it. He was great at free throws, a total inspiration to each of us, although we could sometimes get him at "horse" because of his limp.

Winter was a better time for a guy with a limp. In the evenings Dad set up the card table in front of the TV and painted by number. He had fabulous patience and could spend an entire winter doing a rendition of *Blue Boy* or a landscape of the Grand Canyon. We girls set up the other card table and worked on a jigsaw puzzle with Mom. On Saturday nights we watched Red Skelton. On Sundays it was Ed Sullivan.

Marsha was fifteen when the Beatles were on the *Ed Sullivan Show*. It was 1964. No one else in the house had heard of the group until one Sunday evening when Ed announced their future appearance on that show. Marsha could speak of nothing else. By now she was no longer a priss. During her teenage years she was what you might think of as an airhead. She was obsessed with boys, giggled too loudly and too often, talked on the phone constantly, and, apart from more than her share of pimples, hadn't a care in the world.

She was totally beside herself with the upcoming appearance of the Beatles on television. She had the countdown marked on the calendar, crossing the days off for the two weeks before. She tried to enlist Cheryl and me in her enthusiasm: "They're from England! They're fabulously talented musicians! They're so cute!" We would have nothing of it, our indifference a great affront to her.

On Beatle Sunday, Marsha bathed early, put on her best baby-doll pajamas, and paced around the house until ten minutes before eight, when she landed on the couch, overwrought with anticipation. Cheryl and I were playing Cootie on the rug, Dad was sitting in his chair reading *Time,* and Mom was putting Tracy to bed. At the appointed hour, Ed came on the screen and announced that he had "a really big shew tonight." Then our family saw something we had never seen before. The cameras panned the audience, which was chock full of teenage girls who, when Ed mentioned the word "shew," collectively exploded into what was my first encounter with mass hysteria.

I remember the awe I felt when the curtain rose, revealing those young men called the Beatles. They looked nice enough, in spite of their hair, a length we had never witnessed on men before. We mostly couldn't believe the audience—girls collapsing into the arms of girlfriends next to them, others shivering and shaking as they moved in a trance toward the stage, only to be grabbed by security guards, who led them back to their seats. Other girls were just standing and shrieking, paralyzed by the moment. Throughout it all, those boys on stage just played and smiled, bopping up and down a bit.

About halfway through the first song, we became aware that Marsha was writhing and whimpering on the couch, a noise coming from deep inside her, as if she were possessed by something. It got worse until finally she squealed, "Oooo," in a high falsetto. "Oh, shush!" Dad barked, no longer able to contain himself.

Like Ed's audience, Marsha, too, was out of control. She was incapable of containing what she was feeling then, a cathartic tribute to the power of these long-haired young men in matching suits. She rolled around the couch making little peeping noises until the last chord of "Please Please Me" had faded, the camera had panned the audience one more time, and an ad for White Tornado splashed across the screen. Marsha then lay on the couch, exhausted, staring at the ceiling. Finally she gathered enough energy to pull herself to her feet and head upstairs to her room, where she could meditate on the Beatles' posters that covered her walls.

That appearance on the *Ed Sullivan Show*, and my sister's subsequent behavior, forever changed our views of entertainment and Marsha. Marsha has always had the ability to become totally swept away by things. I guess we're lucky she never encountered the Moonies or Jim Jones, although her Amway stint made us all uncomfortable a few years ago.

Later, years after that Ed Sullivan night, Cheryl and I tried to call up the same outrageous enthusiasm when we attended a Three Dog Night concert or saw Chad and Jeremy at the Spokane Coliseum, but we never succeeded in finding in ourselves the utter abandon Marsha felt. I wish we had a movie of that night in our living room in February 1964 as some sort of testimony to the power of experiencing something so much bigger than we'd hitherto known.

Television was extremely important to my sisters and me, as well as to the rest of the family. Our black-and-white RCA in its wooden cabinet was one of our most treasured possessions, a way to transport us to the world outside of our Palouse. The men from U.N.C.L.E. opened our eyes to the world of international espionage and demonstrated how well Russians and Americans, in the embodiment of Illya Kuruyakin and Napoleon Solo, could work together if given half a chance. The Mod Squad inspired us to dedicate our futures to law enforcement, as well as strive to have hair as straight as Julie's and have a black friend like Linc, the latter goal a challenge if we stayed in the Palouse. Both Dr. Kildare and Ben Casey were instrumental in Marsha's future nursing career. She preferred Dr. Kildare, but both Cheryl and I were wild for Dr. Casey, his surly seriousness contributing, I'm sure, to my adult taste for surly bad boys. We longed for a home life that emulated the Stone family's on the *Donna Reed Show* or perhaps Fred MacMurray's family on *My Three Sons*. Even though the boys had tragically lost their mother, everyone seemed so happy all the time. I think we were probably, in our own adolescent way, just longing to be close to Paul Peterson or Tim Considine or Don Grady, those teenage heartthrobs of the time.

I don't remember my parents watching a lot of television, but they both loved Jack Benny and Red Skelton. My father was crazy for a show called *That Was the Week That Was*, with David Frost as the host. It was a news satire show that held nothing sacred. My first exposure to Buck Henry and the songs of Tom Lehrer came from this show. I'm sure it was ahead of its time and probably why the program only lasted for one year on the air. I didn't really get the jokes, but I remember my dad laughing so hard and my mother being a little scandalized by them.

The only show I remember my dad laughing equally hard at was *Laugh-In*, a show our whole family enjoyed together in the late sixties. I don't remember whether it was Goldie Hawn or Judy Carne who was the Sock-It-to-Me girl, and I can easily picture Dan Rowan and Dick Martin, the hosts, although I can't remember now which was which.

Ed Sullivan treated us to the Beatles and the Rolling Stones. "Hullabaloo" showed us Herman's Hermits and Sonny and Cher. The Monkees, thank God, had their own television show, an early-evening event that left us on edge for the rest of the night. We were so smitten with those guys.

The Monkees went off the air in 1968, a year before Woodstock. Times were changing. The stakes started getting higher, the causes more serious, and the concerns about war and justice erasing any television shows as shallowly fulfilling as the capers of Mickey, Davy, Peter, and Mike. At least for me anyway. At fifteen I had discovered a copy of *Soul on Ice*, Eldridge Cleaver's book on black power, in the Liberty High School library, and that's when everything changed for me. Lord knows how that book ended up in our tiny high-school library, not to mention how it could speak to a white wheat farmer's daughter. Probably some misguided librarian had ordered the book from a "recently published" list because it sounded like a spiritual quest in the Arctic or something otherwise inspirational for young minds.

Six years behind me, Tracy still had years of shallow television watching to fill her time. She grew up with *Donny and Marie, The Brady Bunch,* and *Happy Days,* hardly pieces of work designed for heavy thinking. Her walls were covered with posters of the Fonz, an infatuation none of her older sisters could get a handle on, no matter how hard we tried.

Poor Tracy was too young to participate in her sisters' pubescent fantasies about sixties TV stars. When *Man from U.N.C.L.E.* went off the air in 1968, she was only six years old, never having once felt that wave of fear and concern for Illya, as he risked his life for the safety of mankind. It was Illya we most worried about, so sexy and vulnerable was he in his little Russian, blond way. We were wild for David McCallum, the actor who portrayed Illya. Gail Dennie was actually a member of his fan club. She had ordered pencils inscribed with "David McCallum: Your Man from U.N.C.L.E." from the club. She never, and I mean never, let anyone borrow one of those pencils, for fear of losing them. They were black with gold lettering and always sharpened to a perfectly pointy tip. I'm pretty sure she never even wrote with them.

One thing that we didn't all do exactly the same was attend church. Although we Lutherans were definitely in the majority, the Brewers were Presbyterians and the Martinsons went to the Spangle Community Church. We Hein girls didn't exactly classify the latter as a bona fide church, since our definition of church demanded a denominational definer. Because Grandpa Keller had been the pastor at the Lutheran church in Fairfield, we thought we knew a thing or two about what could or could not be considered a church.

Except in August, Sunday was definitely our day of rest. In our neighborhood, we Lutherans had the understanding that Presbyterians, while basically nice people, were not as deeply religious as we were. We thought of their church, on the other side of town, as more a clubhouse than a real place of worship. The other church was Seventh Day Adventist, located about a block down the hill from the Lutheran church. The Adventists made us nervous. They had church on Saturdays and didn't eat meat. Not only that, they didn't attend our school and rarely associated with any of the townspeople. The only Seventh Day Adventists we really knew were the town doctors—Dr. Hart, Dr. Shirley, and Dr. Francis Theil, and their nurse, Sunshine.

Dr. Shirley and Dr. Francis were married to each other, so shared the same last name, although I was almost an adult before I realized that "Shirley" was my doctor's first name, not her last. All were devout Seventh Day Adventists, for which we forgave them, because they did bring us into the world and saved our lives probably more than once. They were always trying to get us to eat more vitamins and buy those Loma Linda lentil burger boxes the grocery store kept in stock. An evangelical crusade for vegetarianism must have been frustrating in our beef cattle country.

The Academy was the boarding school for Seventh Day Adventists. This was where their children went to school, accompanied by a whole cadre of others whose parents shipped them in from God-knows-where, to live and learn during their high-school years in Palouse isolation. We local non-Adventists called them "the cabbage eaters" and the Academy we called "the cabbage patch." Whenever we drove by, which was pretty often because the road to Spangle cut right through the Academy, we would gawk at those prim-looking children, pale-faced and peaked from a lack of red meat.

I got to know only one cabbage eater well and that wasn't until I was in high school. Jack Eberhardt had gotten kicked out of the Academy for smoking. We're not talking suspended for a few days, we're talking ejected forever. Those Adventists didn't kid around with such high crimes as nicotine consumption. By the time I was in high school, I thought anyone who smoked was okay by me. Jack's reputation, combined with his good looks, propelled me to immediately accept him into my circle. He was my date for the Christmas Ball. I remember with fondness kissing and smoking on the far side of the Grange Hall parking lot during a band break.

Jack pledged two goals in life: never to set foot on Academy property again, and to become an actor. The last time I laid eyes on him was a few years after I graduated from high school. I was in the furniture department of a Sears Roebuck store in the middle of the day killing time. As I tested one of those motorized BarcaLoungers, I started watching the television in front of it, and there he was, Jack Eberhardt, Bachelor Number Three on *The Dating Game*. I watched the rest of the show, laughing aloud at what I considered to be Jack's outrageous answers, answers that clearly established him as far wittier than either Bachelor Number One or Number Two. That gal made a mistake, not choosing him, I thought, remembering our kissing at the Christmas Ball.

The half-church in town was Kingdom Hall, where the Jehovah's Witnesses met. It was a storefront on Main Street with a windowless door and drawn curtains. I never knew a Jehovah's Witness, didn't know of anyone in town who was a Jehovah's Witness, and never heard any sounds coming from their Kingdom Hall. There was something very suspicious, if not downright sinister, about the place. When I imagined what a Jehovah's Witness might look like, I always pictured a man wearing Amish-like clothing who never smiled or had any fun.

One reason the Lutheran church was probably the best, in our opinion, was that it sat on the highest hill in town. "Nearer My God to Thee" was how we looked at it, whatever that meant. In the late 1800s the parishioners built this tall, red-brick building in the classic church design: big basement, giant steeple, and a wide, sweeping stairway leading up to the front doors. In the 1950s a wing was added called the Overflow, a place for people to sit at Christmas services or at a baptism of a baby from a large family. From the front the church looks perfect, with its wide concrete steps leading up to the double metal doors. The steeple looms tall above the roof, with a casing that houses the church bell and a cross. The parsonage sits a few yards away to the left, built from the same red brick.

We knew that church inside and out—from the narthex, a church word for entryway, to the sacristy, that secret room to the side of the altar where the pastor sat and waited while the organist played hymns as people filed into the service. The sacristy was a tiny room with only a chair and a shelf for the communion wafers. Not many children had been in that room, but we had special privileges because of Grandpa Keller.

As newborns, Grandpa would baptize us, with a different aunt or uncle standing up as our godparent, signing those papers and promising to lead us in a good Christian life if Mom and Dad failed to do so. I think my godparent is Uncle Bob, Mom's only brother, who is a bishop now, running the Lutheran church in Idaho, as well as parts of Montana and Washington. With a godparent like that, one would think I'd be a shoo-in for a good Christian life. However, if it counts for anything, among the multitudes of cousins who have resulted from the various unions on my mother's side of the family, I may be the least likely to go to church. This could be considered a bad reflection on Uncle Bob, now that I think about it, although my work with children stricken with cancer has certainly redeemed me a great deal in everyone's eyes.

Our church had the highest ceiling I'd ever seen as a child, second to the high-school gymnasium. It also was the most beautiful sanctuary I had ever seen, if sanctuary is what you call that giant room where everyone worships. It's a lovely word and seems right for a place considered so sacred. Thank God for Lutheran gentility; our stained-glass windows, unlike those in Catholic churches, were not the bloody images of Christ nailed to the cross, enduring unspeakable suffering but, rather, lovely images of Jesus as a shepherd, holding plump, curly-haired babies in his arms and surrounded by little lambs at his feet. My grandpa was not a fire-and-brimstone kind of pastor, either. I was lucky. His message was about love and doing good works for everybody, not just other Lutherans. He even stressed kindness to animals. Only as an aside would he mention what would happen if we didn't behave properly, but in no way did he linger on ugly, unpleasant descriptions of hell.

On weekdays both sets of grandparents often spent afternoons on the farm, especially during the summer, when there was plenty of work for everyone. Grandpa Keller wasn't handy with manual labor, so he would take the children on walks. We would pass through the fields that hadn't been cut, the wheat scratching our faces and arms as we made pathways through it. We were always looking for things. Grandpa made announcements about our discoveries. He would say things like, "See that hole? Do you know who lives in that hole? Mr. Badger lives there. Why don't you say hello to Mr. Badger, but say it softly because he might be napping." So we would whisper hello down the hole, wait for a response, and when it didn't come, keep moving through the fields or the draws until we found something else. We were usually searching for insects or some sign of animal life. Grandpa told us what the creature wanted to say, which was usually just a friendly greeting. Since he was a pastor, we never doubted that he knew.

After Grandpa and Grandma Keller moved to Oregon, Pastor and Mrs. Willman came to town with a son, Mark, who pulled my braids and made me cranky. Generally speaking, preachers' children are hard to get along with. The girls tend to be prissies and the boys are bullies. I suppose it makes sense that a normal, well-adjusted childhood would be difficult. Growing up in a series of parsonages with a father everyone called Pastor Willman and a mother everyone called "the pastor's wife" might cause one to break out a bit. In Mark's case it was in merciless teasing and insistent braid pulling.

To be a preacher's kid is to live in a fishbowl. Congregation members expect their preacher's parenting skills to be a model of divine intervention, with the resulting behavior to show for it. This is rarely the case. Every nonpreacher's kid knows it, and every nonpreacher's kid in the congregation is attuned to the slightest transgression, always ready to chant, "Guess what Mark Willman did?"

Rarely would any of Mark's friends want to spend the night at his house, which wasn't even called his house but, rather, "the parsonage." It was possible that one might see Pastor Willman in a T-shirt, or shaving, or, worse yet, coming out of the bathroom with a magazine, or even the Bible, in his hands. One could witness something so totally unexpected and unpastoral that Sunday sermons would be changed forever. It was too risky. Besides, the church, a few yards away, was a welcoming place during the daylight but very creepy after dark.

Even being a pastor's granddaughter had its drawbacks. But in hindsight, my sympathies are much more with Mark. There were, however, some benefits for him. With few boys in the Sunday school, and Mark being who he was, he was assured every year the role of Joseph in the Christmas program. Joseph sat on a folding chair by the manger for the entire Christmas service, which sounds a little boring, and was, but it sure beat memorizing Bible verses and delivering them in front of a packed church. It bothered me that the Mark who tormented me would be cast as the earthly father of Jesus, but I comforted myself with the knowledge that at least he wasn't cast as God the Father.

I don't recall ever being the Virgin Mary, not even once. That role usually went to Kathleen, a girl one year younger than me who lived in Rockford, a few miles north of Fairfield. Everybody knew she couldn't memorize a Bible verse if it were her ticket into heaven. Mary and Joseph were nonspeaking roles, every year played exactly the same way. Kathleen and Mark, otherwise known as Mary and Joseph, walked slowly down the aisle toward the altar and the wooden manger. Kathleen held a plastic doll, which a Sunday school teacher had wrapped in swaddling clothes, which, I guess, were flannel baby blankets. Kathleen deposited the doll in the manger, and, throughout the whole program, Kathleen looked maternally at the doll while Mark sat on his chair, leaning on the broom handle he carried as a staff. The rest of us took turns walking to the front of the church and rattling off verses from the first part of the New Testament. We shifted from foot to foot, anticipating the peanuts and candies that the ushers passed out at the end of the service. Only after all the candles had been lit, the lights had been dimmed, the congregation had sung "Silent Night," and the church bells had sounded their last deep, baritone peal would we get those brown lunch bags full of treats.

The Christmas program was the same every year. It happened on Christmas Eve and, regardless of the weather, we drove to town for it, dressed to the nines in something velveteen and new. We wore our winter dress coats and our snowboots, keeping our patent leather shoes tucked in plastic bags for later when we were in the church. Dad played Christmas music on the radio while Mom sang along. We children looked out the windows, going over and over in our minds the Bible verses we were about to deliver, silently moving our lips.

Grandpa and Grandma Keller had given each of us our own Bible with our name embossed in gold on the front, next to the date of our birth. Mine had a white leather cover that zipped open and closed. It had a red satin bookmark and a place where I could write my family tree under the heading "Everyone Is a Child of God."

The Hein girls were known for their ability to memorize, a skill that haunted us throughout our childhoods. Whenever a Bible verse was in need of recitation, Pastor Willman knew he could count on a Hein. Whether it was a shorty from Luke or a whole paragraph from Leviticus, we could get it word perfect in no time.

At Christmas I almost always got John, chapter 3, verse 16: "For God so loved the world that he gave his only begotten son that whosoever believeth in him should not perish but have eternal life." I must have recited that Bible verse a thousand times and never even once wondered what the word *begotten* meant. It fell into the category of "swaddling clothes"—just go with it. I was a memorizing mynah bird. I could do a whole psalm without even a glimmer of curiosity about the content. I didn't care. I would just memorize it, standing for hours in front of Mom's big mirror in her bedroom, holding my hands in a reverential way, looking sincerely out into the audience of my imagination, delivering these words from God as if He Himself were speaking through me. I intended every year to transport this countenance out of the bedroom and into the church, where I would recite my verse with confidence and candor, making eye contact with as many parishioners as the length of my Bible verses would allow.

It never happened, though, in quite that way. Every year, there I was in my newest velveteen on Christmas Eve in front of the whole congregation, desperate to do it right. It was snowing outside, and the church was lit by candlelight. The pine boughs on the windowsills and a packed church gave a kind of sweet-smelling warmth to the air.

At home we'd finished trimming the tree and had eaten our Christmas chowder before coming to town for church. I knew after we got home we would light the angel chimes one more time. In the candlelight given off by the lit Advent wreath hanging in the living room, with the angel chimes *ding-ding-dinging* in the background, as every Sunday night for the past month, we would sing Christmas songs. Then we opened a window on the Advent calendar and listened to Mom read a couple of Bible verses. That all lay ahead of me once I had delivered my verse to the congregation, the very same verse that I had practiced so assiduously for the past month in front of the mirror.

It happened the same every year. Everybody in the Sunday school stood around Mary and Joseph and the plastic Baby Jesus for the reading of the Christmas story. Then the little kids sang "Jesus Loves Me" and "Away in a Manger," and then there was a

short sermon from Pastor Willman. When the whole congregation sang "The First Noel," we reciters all knew the time was
coming. I began to worry about a tickle in my throat. I began to
forget the beginning words of my verse. I began to wonder if my
shoe had come unbuckled, if my slip was showing, or if my
shaking knee would allow me to walk down the two steps to
the space in front where I was to recite. I began to wonder if the
sound of my heart in my ears was audible to the congregation
and if I should try to speak louder to be heard over that heart. But
I mostly worried if I would be able to speak at all.

By the time the last chorus had died down I was in a trance
of worry that led me down the two steps and into the middle
between the pews. Somehow, with my eyes plastered to the clock
that hung on the balcony in the back, I said those verses, word for
word in a monotone voice without a pause, like a robot child of
God. I tacked on the source at the end, "John 3:16," without so
much as taking a breath, showing no vocal inflection to imply it
wasn't exactly part of the message. Then I moved back into my
place while Gail moved to the vacant spot to recite her verse, and
then Cheryl, and then Susie, and so on, until we had each gotten
out our practiced parts in low mumbles or stutters, some with
Margie, the Sunday school teacher in the front pew, mouthing the
whole thing in exaggerated lip movements.

The service was over after the congregation sang "Silent
Night." As they started the final chorus, someone in the balcony
rang the church bells. Then the acolytes put out the altar candles,
and Pastor Willman said the benediction: "May the Lord bless us
and keep us. May He make His face to shine upon us, and be
gracious unto us. May He lift up his countenance upon us and
grant us peace. In the name of the Father, and of the Son, and of
the Holy Spirit. Amen." Pastor Willman would then glide down
the aisle in his robes as Elsie, the organist, hit the keys. We
children filed out, filled with relief that the verses were over,

although disappointed that, yet again, we had failed to deliver our verses with conviction.

Before leaving town we piled into the car, which Dad had already warmed up, and drove around Fairfield looking at the Christmas decorations. They were exactly the same every year. Dr. Hart's were the best, with his wooden, life-size replicas of Mary and Joseph in his driveway. He covered his whole house with blinking, colored lights, the kind with the air bubbles burping to the top over and over again. The Daniels with their laughing Santa and the Schoenhauers with their Rudolph on the roof were close rivals.

The church was far from just another building in town to us. Grandpa married Mom and Dad there, not to mention Uncle Bob and Aunt Betty, and my sisters Cheryl and Tracy. We all were baptized and confirmed there. It was also the site for both Grandpa and Grandma Keller's funerals.

I was a teenager when Grandma hit ten years at the Fairfield nursing home. Her Alzheimer's had caused her and Grandpa Keller to return from Oregon. It would rob her of her speech, her walk, her reason, and just about everything else except for eye contact and the love my grandfather showed her every day, feeding her, talking to her, and telling her stories. He called her "Mom" and insisted that she be gotten out of bed every day, have her hair done, and her teeth put in. He gave her a cloth to hold in her hands so her fiddling wouldn't rub the skin away. For Grandpa, her Alzheimer's was sad and inconvenient, but he didn't lose any sleep over it, because from his perspective, her illness was remarkably temporary in the scope of the eternity they would spend together. For the rest of us, as convinced as many members of the family were of eternal life, her illness was a horrible tragedy. Grandma Keller was a powerhouse, a brusque woman with strong opinions and a loud voice, traits she passed on to every female she birthed.

She met my grandfather in Sawyer, North Dakota, in 1925. He had arrived at her small town to be the parish minister. He lived in a former grain elevator during this, his first call. She was the church organist, in charge of this magnificent musical machine that sat in the little white church high on a hill overlooking a huge North Dakota landscape. North Dakota is endless. The sky bubbles with every type of cloud ever pictured in a sixth-grade science textbook, each one rolling around with that wind that never ceases. For hundreds of miles, the view from anywhere is uninterrupted by anything as distracting as a tree or a mountain or a building. Rolling hills, wheat, wind. That is all and that is plenty.

Grandpa's first congregation was mostly German. He preached the early service in that language. Grandma sang as she played the organ to accompany herself and the congregation. The two of them repeated the message and the hymns in English immediately after the German service. They were first-generation Americans and knew both languages.

Grandpa always said he loved her immediately, and I believed him. They married, had five children, and moved through their lives together, he preaching and she playing the organ. Throughout the Depression his pay was often jars of jam and ears of corn that she would miraculously turn into meals. This may not have saved all her children from rickets, but it kept them alive. The song "Stand By Your Man" could have been written by my grandmother as she followed hers from parish to parish, the children in tow, living in poverty, teaching piano, heading up endless Women of the Church committees, and insisting, above everything else, that her children love God and be well educated. She herself had an eighth-grade education, but that didn't stop her from saving her piano money for her children's college tuition. Every one of them was born on the plains of North Dakota into a poverty of sorts. Grandma expected each to be much bigger than their environment assumed them to be. If love and brains are the measuring stick, each one of them is.

Two are ministers: Aunt Karen, her gender unusual in the profession, and Uncle Bob, previously described as the family bishop. Two others are nurses, one of them also a pastor's wife who manages to stand by her man and be a nurse, a kind of double duty understood only by the wives of pastors and presidents. The other nurse found money, in a resistant town, to start a hospice to help people die in peace. My mother is a teacher.

Grandma was a bit intimidating to her grandchildren. She wasn't a big, soft grandma with a lap you wanted to snuggle into while munching on some warm, freshly baked treat. This particular Grandma specialized in sauerkraut, not cookies. Grandma herself was a tough cookie who had a gravelly smoker's voice, even though she had never smoked. That voice could often be heard prodding along whoever was her latest target. She berated us for not practicing our piano, not getting straight A's, and not keeping our clothes clean. She was a mean "Make a Million" player, and no one could beat her at Scrabble. She was a fabulous cook, never once buying a ready-made loaf of bread, and regularly filled the table with sauerbraten, knoëdel, and *apfelkuchen*, not to mention meat loaf and mashed potatoes.

When Grandma's illness slowly sucked her into oblivion, none of us really believed it. We made our token visits, looking into her eyes, trying to determine if she could recognize us, as if somehow that would make a difference in how sad we felt. We made idiotic one-way conversation as we told her what we had done that day, what we were doing with our lives, or asked her questions about what she had had for lunch, as if she could answer or do anything except fidget with her cloth.

In 1990 Grandma Keller died. It was hard to feel genuinely sad. We all arrived for the funeral at a moment's notice: her children, her grandchildren, and her great-grandchildren coming from all over the Pacific Northwest, if not farther, to the church in Fairfield. Uncle Bob, who garners all of our respect for his thoughtfulness and cool head, told us how to walk in and where to sit. The front pews were reserved for us, and we were to file in through the Overflow, while everyone else came through the front door.

I sneaked into the church earlier to have a peek at Grandma, who lay in her coffin by the altar. I thought this gesture important, remembering what Mom had said years before about David Brewer and never believing he was truly dead, which, in fact, was sort of true. The Fairfield mortician was on vacation when Grandma died. His replacement, a man from another town who was unknown to us, had done a horrific job on what was left of my grandma. I'm sure that an eighty-five-year-old dead woman who had Alzheimer's for ten years was not an easy project, but an inch of industrial-strength pancake makeup wasn't the ticket. I left immediately. Distracted by this image, I filed into the church, my head down, my heart sad, and my mind puzzled by the body in the box.

My grandma had not spoken to anyone in more than ten years and hadn't lived consciously in Fairfield for thirty. Funeral attendance predictably could have been slight. It was only when the first hymn began that I knew her family was not alone. What I hadn't noticed when I walked in was that the church was full. Pastors, their wives, former parishioners, adults who had once been her piano students, and adults who had not—all filled the church. People had driven from far away to be there, and the place was packed. I was startled to see so many people from her life whom I had never seen or even heard of before. Where did they come from? At that very moment, as the first verse took off, it seemed that half of them knew how to sing harmony and the other half were strong leads, and "Oh, Take My Hand, Dear

Father" never sounded so majestic. It was as if every voice who had ever known my grandma, this organist from the plains of North Dakota, her shadow lying in a box in front of us, was singing behind me. I couldn't even peep a note, nor could anyone else in our family. Grandma had taught each and every one of us that hymn and now we were speechless. In that second she no longer was "Grandma with Alzheimer's" but had deconstructed into Grandma Keller, alive in our memories as the real woman she had been for the better part of the century. "This one's for you, Grandma," I thought, as my cousin Dianne, who was the church organist, hit the chords for the second verse.

It was in 1967 that Barbara Martinson was diagnosed with lupus, a weird disease that makes a person's body allergic to itself, at least that's how one book described it to me, when I decided to do a school report on the disease after Barbara's diagnosis. With lupus, the victim's body overreacts to some outside stimulus, such as something environmental or some infectious agent, and then starts making too many antibodies, which somehow get directed against the person's body tissue. Women between fifteen and forty-five are the most common sufferers of this disease, which often goes undetected at first since the symptoms are fatigue, achiness, stiffness, and low-grade fevers. It's like having the flu, at first anyway.

For Barbara, it seemed like an especially cruel disease to get, because both she and Joel were such fun-loving people. The worst part was that lupus isn't curable. Barbara would be tired and achy for the rest of her life. The doctors were pretty certain, though, that if she took care of herself, she wouldn't die from it, just feel tired all the time. They assured her that fewer than half the people that get lupus experience any organ damage, although it was true that if it did get hold of her kidneys or liver, she could be done for. I remember thinking at the time that "fewer than half" wasn't really my idea of a good statistic, but she seemed comforted by it. She was only twenty-six, although that didn't seem so young to me at the time.

When Barbara was diagnosed, *Mission Impossible* was on television, and one of the starring actors was a man named Peter Lupus. The disease wasn't named after him, of course, but somehow watching the program after her diagnosis made me both more confident and more worried for Barbara—more confident because there were always solutions on *Mission Impossible*, no matter how complicated the issues were, but also, anything having to do with *Mission Impossible* was always very ominous and deadly. I quit watching the show during this time as I grappled with my mixed feelings.

I was in high school and always in desperate need of money. Before Barbara's diagnosis I made a fair amount baby-sitting the Martinson kids. Because I was often grounded for recalcitrant behavior during my teen years, the Martinsons could count on me to be available most Saturday nights. After Barbara got lupus, she hired me on weekends, mostly to come over and do housework. Because she asked me to do the ironing, including Joel's shirts, she paid me more than she did for simple baby-sitting. Barbara seemed unaware that there was nothing more difficult than baby-sitting the children she had raised, and also unaware that I was the world's worst ironer. She was just so tuckered out all the time that it really didn't matter how poorly I kept house. I'm sure Joel didn't notice either. He had his hands full with a sick wife and three young children.

I was away at college when Barbara died. In spite of what her doctors said, and in spite of the fact that she took care of herself, she died anyway. She just got more and more tired until she was gone. Or maybe she did eventually have liver and kidney failure, I'm not sure, but there was just nothing they could do about it. She wasn't more than thirty-one years old when she went.

Mona Zehm died three years later.

For eight years after her surgery she lived, her speech slurred and parts of her body slowly curling up into her metal braces. Then, like a racehorse out of the gate, her tumor returned from whatever place it had been lurking for all those years. Leonard kept her at home as long as he could. One Sunday morning he accidentally dropped her as he was lowering her into the bathtub. It was the old-style tub with the sloping back, and, luckily, he had overfilled it that morning. The water helped catch her fall, but she still got a resounding smack on the head. "It's okay," she told him, "no need to feel so bad about it." But Leonard did, knowing that her ravaged condition, which was steadily worsening, was eating away at his own strength, making him too shaky and unreliable to take care of her in the way he should. Together they decided that Mona needed to go to the nursing home. Leonard bundled her into the car. As they drove up on the other side of Hangman Creek, Mona noticed where a farmer was going round a field on a combine. "Isn't that beautiful?" she said. "What a shame I'll never see it again."

Two weeks later she died. They called Leonard to come quickly from home. He had just left Mona's side briefly to return there for a shower. It left all of us in the neighborhood with mixed feelings of relief and sadness.

I have two pictures of Mona in my mind. One is of her clutching Leonard's arm as they make their way down the aisle at church. Pre-surgery and post- they sat in the fifth pew from the front. They attended every Sunday, and every Sunday they moved down the aisle in the same way, Mona smiling her greeting to everyone and Leonard looking rather courtly with his elbow crooked, holding her up through their slow processional walk down the aisle. They were like a surreal royal family. Even during harvest Leonard brought her to church—one of the few men under eighty years old in August attendance.

Another memory is a snapshot my mind has been gracious enough to hold. It is a memory of Mona before she was ill. She and my mother are hanging on either side of our backyard fence, and they are laughing at something. The joke is so good they have to hold onto the fence to keep from falling over. It must have been June because the lilacs are in bloom, lavender and white in the background. Polly is clutching Mona's leg, a toddler going along for the ride of this joke, whatever it was. The two women were good friends, and I think now they were both rather beautiful, especially when laughing, although Mona's beauty was long gone by the time I considered those sorts of things.

I don't believe that Dad and Mona ever had a significant conversation about their shared illnesses, although our fields were back to back, and cancer, even in the fifteen years between each's diagnosis, remained an anomaly, a bizarre twist of fate. People there, though, don't like to speak much about unhappy topics if they can avoid them.

* * * * *

I remember learning of Jimmy's cancer when I came into the kitchen one afternoon, home on a break from college. Mom and Dad sat across the Formica table from each other. Mom drummed her fingers against her coffee cup, and Dad looked out the window at a magpie as if he were an ornithologist scrutinizing its features. They weren't speaking to each other, although I knew it wasn't out of anger. It was nothing more than that one sentence Mom said—"Jimmy has cancer"—and everything changed.

It must have been a combination of my advancing age and the tone in her voice that led to a strange feeling about illness and my neighborhood. I think of this as the moment when I first felt our vulnerability. Jimmy was forty-one years old and had colon cancer. By now, Dr. Hart had taken to sending everyone from Hangman Creek directly to specialists in Spokane whenever there was even a whisper chance of cancer.

People were getting an odd feeling about our neighborhood by then. Dad had had thyroid cancer. Mona Zehm was dead from her brain tumor, David Brewer from leukemia, Barbara Martinson from lupus, and now Jimmy was sick. If we had had a bunch of farms between our farms, it wouldn't have seemed so odd, but, in fact, we didn't. We were one farm right next to another.

Jimmy's news confirmed that I should stop thinking everyone lasted until retirement, although you would have thought with our neighborhood's track record, this would have occurred to me sooner. Before, I assumed people died only after they had passed their piece of land over to the next generation and after they grew to a ripe old age. People needed to wobble year after year down the aisle at church, aging each Sunday before our very eyes, until they finally went to the nursing home up on the hill and didn't even make it down for Christmas services.

It seemed as if almost everybody eventually ended up at The Home, although, as children, when we went there to visit, it seemed an unimaginable consequence of life. We would see everybody's grandparents there when we went caroling or when we went to visit our own grandparents. It was a shockingly sad place with bent-over toothless people who looked only vaguely like the old folks we recognized from a few years earlier in church. Now they were astonishingly old, older than we thought people could ever possibly get. Luckily, we thought, the one thing you could conveniently count on was that the indignity of this condition was temporary. We all promised ourselves to remember them as they last looked, sitting erect in the church pew, not slumped into a parked wheelchair in a stuffy room that smelled vaguely of urine.

The Home symbolized the ending of the life cycle. My Grandpa Keller, who lived there until he died, gets credit for creating The Home. He was a pastor in Fairfield, a fairly young man with several generations of Fairfieldites to minister to, when he decided there should be a place for the elderly in town. Before that, anyone who couldn't die at home went to the city. Grandpa thought that people who had lived their lives looking out over wheat fields and lilac bushes should die looking out over the same thing.

It was Grandpa who spotted the old Morrison mansion up on the hill, an aging remnant from the former sugar-beet king, and persuaded his parish, as well as the Presbyterians and some miscellaneous townspeople, to chip in to start The Home. Augusta Blakely, Mable Stenson, and Earl Stecks were the first to live and die in those upstairs bedrooms, each one with a nurse at his or her side, as well as family members and my grandfather. The plaques made soon after each of their deaths still hang in The Home office, commemorating the first three residents' distinctive way of leaving this life.

The Home is much larger now, ministering to the nursing-care needs of several of the small towns in the area. They tore down the Morrison mansion a number of years ago and replaced it with a single-story sprawl that droops over the top of the hill like an unstarched doily. The center of the doily is an outdoor sitting area. Circled around this are the residents' rooms and the convenience apartments for those who can still cook for themselves.

Grandpa Keller lived in one of those apartments until he was ninety-six years old. Until the last year of his life, he microwaved Lean Cuisines for lunch, ate dinner in the dining hall with the residents who were able to get there, and led the Bible study on Wednesday evenings. There wasn't a lot of discussion during these classes since my grandfather was almost totally deaf, but after seventy years as a minister, he had enough sermons under his belt to keep the study moving. He had a Bible verse for every occasion, delivered them liberally, and happily led an appropriate hymn when the class was over. He didn't care if anyone sang along with him because he couldn't hear them anyway, and he knew, even if the parishioners weren't singing, they were feeling it in their hearts.

Before he died, I asked Grandpa when his happiest years were. "It was really good in my fifties," he yelled at me that day in the nursing home. It was, in fact, the last day I ever saw him alive. "Don't ever smoke," he then said to my teenage nephews, who were with me. "I used to smoke so much that I'd have to keep the sermons short so I could get out to the car for one. I promised Mom if she had a boy I'd stop smoking, and then Bob came along and I did. I had to keep my word. After that the sermons got longer."

Even his last conversation with us was a sermon designed, as always, for our betterment.

It was a blow when Jimmy got his cancer. Even though I considered anything over forty to be close to ancient, I knew it was far too young an age to go to The Home, which seemed what everyone should do prior to death. His wife, Dona, was even younger, still barely in her thirties, a ridiculous age for a potential widow.

Jimmy and Dona started taking more vacations, to Priest River, to the Okanogan, and one time they even took the whole family to Disneyland. Jimmy had four children. The oldest, Greg, was a couple of years younger than me, and the youngest, Jill, was three years younger than Tracy. Sometimes Jimmy was too sick and he canceled the trips. There were also chunks of time when he was just fine, keeping up his farming with help from his friends and neighbors, who kept their eyes on the Hahners' house up there on the hill, watching for any trouble they didn't want to see.

My father went over and helped when he could, but he had his own problems. The limp had dramatically worsened after his Christmastime brain hemorrhage. His balance was weird, his moods could swing from cheerful to abject in minutes, and farming, with its multitude of demands, both physical and mental, had become impossible.

"It's not that he can't do it," my mother explained to us. "It's that he can't do it perfectly." There were certain things that my father liked to do perfectly, and farming was one of them. That's why our farm looked like a postcard. Dad never left anything unpainted, unshoveled, or untidied. We got rid of the chickens because they were too messy, our dogs were purebreds, our cats all short-haired, our horses well groomed, and, years before, he had switched from beef Herefords to registered, polled Herefords, lovely red, hornless animals that, as the most picturesque of cattle, dotted our pastures, making the landscape that much more beautiful. We led them on halters, took them to shows, and auctioned them for lots of money as breeding stock. Our farm machinery was nurtured to avoid the expense of newer models. Dad lined up the machines in the sheds, greased, tuned, and always ready to go. The maintenance of our place demanded daily attention and hours of hard work. With his misplaced balance, poor circulation, and unpredictable moods, it had become an uphill battle.

If Dad couldn't take care of his farm his way, he wouldn't take care of it any longer. He made the decision one day in 1970, when he simply could not get onto the combine, his bad leg refusing to make that giant stretch up to the first rung of the steel ladder that hung off the side.

This was a hurricane occurrence in our otherwise orderly sense of the universe. With no son to ascend to the combine throne, Dad chose Darryl Flaig to farm our fields. He was the younger brother of Dick, the hired man who had turned into Marsha's boyfriend so many years before. Darryl had taken over his parents' house, was farming their fields, and had told my father years before that he wanted to farm our fields. Of course, he had had his stint as a hired man with us, and Dad knew him to be a competent young man.

We all suffered a painful day when Darryl drove our combine down the road to his farm, wanting it closer to him for winter repairs. We bit our lips when he put his motley collection of mixed-breed beef cattle on our creek place. They were surly steers with too much white on them that snarled and shook their horns when we walked by. We didn't say anything when Darryl took down the white-board fences around our pastures after we auctioned the last of our cattle. It was all part of the agreement. Besides, the only thing those fences were holding in now was empty space and the ghosts of favorite horses and pedigreed cattle.

Little by little, Darryl emptied our buildings of drills, sledge-hammers, welders, skeins of rope, hacksaws, anvils, and other things, many of which had been there since Grandpa Hein ran the place. Our barn didn't need to be filled with hay any longer since Darryl wintered the surly steers at his place. Our shed didn't need to house the combine because Darryl did repairs in his shed. Dad moved the tools he wanted to a small workshop down in the basement and let Darryl take the welder, the compressor, the rodweeder, and the manure spreader down the road to his place. Dad sold the saddles, threw away the baling twine, rolled, knotted, and disappeared all the ropes, and burned up the laths that he had long ago retired from the rafters, where Grandpa used them to hold together wet plaster.

We couldn't fight the feeling of utter sadness. The farm had been transformed into something more like an impersonation of a farm. Nothing happened there. Dad swept the outbuildings meticulously clean, and they echoed emptily. He manicured the lawn perfectly, even the part out by the barn where they used to butcher the steers. He weeded the rose beds, edged the sidewalk, and pruned the fruit trees. Then he did these jobs again.

Darryl plowed right up to Dad's vegetable garden and right around the cottonwood tree where we had buried Rockette. It bothered me that he would plow so close to her grave, but I said nothing. Our family adhered to a conspiracy of silence, each of us mourning privately the things that marked the disappearance of our farm as we knew it.

Dad was fifty years old.

Before too long he got a job in the county assessor's office in Spokane reviewing other farmers' tax forms. I had gone away to college and completely missed my father's reluctant and fairly surreal transition from farmer to commuter. Mom reported that he liked a couple of people in his office and that was about it.

In the evenings he swept the empty barns, mowed the lawn with his riding mower, and tended to his newly created vegetable and flower gardens. When we came home from college on vacations, it was peculiar to find him gone every day at exactly the same time and then home every day at exactly the same time.

On the weekends he drove over to Jimmy's and helped him where he could. When Jimmy's cancer was diagnosed, the surgeon removed his colon, but the cancer had already spread to his liver. The chemotherapy Jimmy endured did little to keep it at bay. Dad had known Jimmy all his life, although it was Jimmy's older siblings who had really been Dad's peers during childhood. Jimmy and Dad's friendship had flourished with their back-to-back farms and identical histories. Now they were a couple of middle-aged men made old by serious and wracking illnesses.

Greg, Jimmy's oldest, had gone away to Wenatchee to work on his uncle's orchard and attend community college there. He had decided to take some classes and try another kind of farming before committing himself to the family wheat farm. Six months before Jimmy died, Greg came back to help on their farm, having decided two things: his father needed him and wheat was more for him than apples and cherries.

Greg was twenty-two when his father collapsed in the field. It was their Keevy Quarter, I believe. Greg picked Jimmy up off the wheat stubble and carried his by-then tiny body to the pickup. He drove fast the ten miles to town, stopping every few minutes to desperately give his father CPR, until he gave up this effort and fled toward town. The people across the street at Dodge's Thrift Grocery Store watched Greg barrel down Main Street to screech to a halt in front of the doctor's office. Then, they said, it seemed as if he went into slow motion. Greg leaned his head against the steering wheel for a moment or two. Then, creaking his door ajar slowly, he let it swing wide open before he got out, also slowly. Then he pushed the door shut. Only when it had banged and the echo had stopped did Greg walk deliberately to the other side of the truck in those big, heavy boots that farmers wear. He let Jimmy's door swing open, catching his father's body as it fell out of the truck.

That is when the pandemonium broke as the men in Dodge's Thrift ran across the street to help Greg. The women stood in small groups watching, with their hands to their mouths and their voices stuck in their throats. Of the women, only Claudia Dennie moved, heading for the telephone at the grocery store to call Dona.

* * * * *

Mom was convinced that her daughters must get a four-year college degree so that we would be able to get a good job and not be dependent on husbands, who could die on us at any moment.

My mother has been sure for the past forty years that at any moment my father is going to keel over and die, leaving her a farm widow and heiress to a fleet of John Deere farm equipment. She thought it was her motherly duty to insist that her daughters get at least a bachelor's degree and that we go to the other side of the state to get it. This would serve to broaden our horizons, giving us, in her mind, the most options. If we got a college degree and then decided to return to Fairfield to marry a farmer, that was one thing. But too many local girls never considered anything seriously except marrying a farmer. A girl would put in her token two years at the local community college, while her high-school boyfriend was spending his days disking fields and spreading manure. Around and around the fields he would go while waiting for Patty or Sherry to finish up that two-year course in dental hygiene. Then they married and set up housekeeping in a trailer or manufactured home, installed on some piece of land near the boy's parents. The couple would eventually move into the big house, but only after his parents retired to an apartment in the city.

Whether my parents knew it or not at the time, by sending us off to college on the other side of the state they were kissing good-bye the possibility that one of us would marry a farmer and take over the land. There may have been some vague hope that, after knowing what life out there was like, we would choose the farm, but as each daughter married a city boy, went to graduate school, and/or moved far away, it became clearer and clearer that Dad's successor was nothing more than an abstract idea around our farm.

Darryl was plowing away at our fields, but he lived down the road and was his father's successor. He lived in his family's home with his wife and children and would never have a relationship with any part of our farm except for the fields. His base of operation was the Flaig Place. His advice came from his own father. My father was not so much his mentor as his landlord.

24 A son is born into a farm family. He is raised on the land. He feeds cats at five years of age, horses at eight, opens up bales with a cutter at ten, and drives truck at twelve. For the next five years, he learns to run every piece of farm equipment, repair it, and store it. He also learns how to dehorn cattle, as well as castrate them, and to worm horses, behead chickens, and butcher steers. During this time he might work for a neighbor farmer, expanding his knowledge to include milking cows, raising pigs, or experimenting with a different variety of grass seed. When he finishes with high school, he might go to community college for a brief time, taking classes in vocational agriculture or bookkeeping. The classes may not have a great deal of practical application, however, and the general rule is that after two years at community college in the city, he returns home to farm with his father.

Over the next few years the picture is finely tuned. He and his father farm side by side, while the son learns the subtleties of the profession. He might live with his parents, or, when he decides to add a wife to the equation, make some other arrangement. Often the addition of a wife will cause his parents to move to a smaller house nearby, and the son, plus family, will move into the big house. The father continues to farm alongside the son, although gradual shifts in power are taking place. Each arrangement for division of profit is private and relative to the family situation.

At some point, the father turns the farm over to the son. It happens slowly and quite surely. No one may actually know the transfer is complete until the parents die, although often the transition has an earlier, distinct beginning point.

When Greg Hahner was twenty-two his father died, only three months after Greg had returned from the Wenatchee orchards and his brief time at the community college. He was exploring career options. He could have been a county agent, working in one of the extension offices, giving farmers advice on new high-yield grass seed. He could have taken some practical business courses and bought into a John Deere dealership in Tekoa or maybe Rosalia. He could have taught shop at the high school, moonlighting as the Future Farmers of America faculty adviser. He could have worked at the Grain Growers Association, either in front of, or behind, the desk. Then one day he realized they were all a variation on one theme—farming. That was what he really wanted to do.

A few months later his father was dead, and Greg was living on the hill with his mother and his youngest sibling, Jill. Bryan and Ronda had already gone away to learn their careers.

Jimmy died in the middle of March, not the best time of year for a farmer to go, with spring planting on the horizon. The neighbors rallied to help Greg get the seed in, telling him through his grief what to plant and where to plant it. Greg obediently went round and round those fields, seeding, fertilizing, disking, and taking in the new meaning of the landscape.

While he was only a couple of years behind me in school and we had ridden the same school bus for our entire school careers, I really hardly knew him. Each of us children on that bus had distinct personalities. I was noisy, Cheryl was bookwormish, Jan was giggly, Lynn was bossy, and Greg was quiet.

The ride was forty-five minutes each way over narrow gravel roads that twisted through the fields. In the winter, when the county didn't plow all the roads, the route changed. Then it took an hour and fifteen minutes, the bus hugging the main road, dipping up gravel driveways if the county had them plowed, or dropping the children off at the ends of the roads if a parent was waiting. We were always the last off, because our farm hung on the fringes of the school district. Our road was usually snowed in, but we had permission to walk alone up the half-mile drive. Mom had hot chocolate waiting on the stove. Over our mugs we reported on whose driveways the county had plowed and which roads were still closed.

On the bus, Greg Hahner and Bruce Dennie always claimed the back-row seats across from each other. They were best friends, although rarely did either one of them utter a word. They based their friendship on a mutual pact of silence and the fact that they were cousins. Whenever we described a quiet person, we would say, "Quiet, you know, like Greg Hahner or Bruce Dennie."

I don't remember liking or disliking Greg. He was just Greg, Jimmy and Dona's son. He was in my Sunday school class, as well as in Vacation Bible School, but, just as on the school bus, he and Bruce stuck pretty much to themselves. They were the same age and lived within a mile of each other, the Dennies' farm being another mile east from the Hahner farm.

The only time I can remember hearing Greg's child voice publicly was on his Confirmation day. The Confirmation class that year consisted of Greg, Bruce, and Sylvia Lytle. Each had to recite a Bible verse and either the Apostles' or the Nicene Creed in front of the congregation. I don't know which was more difficult for Bruce and Greg: to walk down the aisle in those white choir robes that resembled girls' nightgowns, or to recite in a voice that was loud enough for even those in the balcony to hear.

I remember Pastor Willman asking in his regal, pastoral way, "And how can we be sure God will keep us from evil?"— a cue to the three to put Psalms 23, verse 4, on the tips of their tongues. After a pregnant pause the pastor would say, "Greg," and Greg's soft voice would push the barely audible words back toward those of us in the pews: "Yea, though I walk through the valley of the shadow of death, I will fear no evil; for Thou art with me." Pastor Willman made it a point of not asking Greg and Bruce to recite one right after the other. The span of barely intelligible mumbling would be too long.

After one mumbled Bible verse from Greg or Bruce, the pastor enlivened the moment with something from Sylvia, who, unlike her colleagues, had an operatic set of lungs and a personality to match. "I believe in God the Father Almighty, Maker of Heaven and Earth," Sylvia would boom out over the congregation, waking Willy Myers, who was snoozing in the balcony. Sylvia sincerely believed she was bringing the word of God to this sleepy congregation and delivered her creed accordingly. While Greg and Bruce had all they could do just to say a Bible verse, Sylvia recognized the perfect opportunity to connect spiritually with the flock in front of her. She spoke loudly, she made eye contact, and she finished each recitation with a small, wizened smile on her face.

Sylvia has since left the Lutheran church. I think it was a bit too reserved for her fevered tastes, its stoic members not sufficiently appreciative of her gifts. She still lives in the area, having married her father's hired man, but now drives over the state line to a church near Worley where it is rumored that people speak in tongues.

Our school district covers an area of 350 square miles, providing for the educational needs of children from six towns and the farms in between. Until a few years ago, there was one grade school in Fairfield and one in Spangle. The one high school, a building that sprawls out over a country meadow in the middle of the district, is large enough for the classrooms, gymnasium, football field, and one locker each for the 150 freshmen through seniors who attend it. At least that's how it was when I graduated. It might have a few more students now, and the grade schools have been consolidated into one that sits big and new behind the high school.

Sometime between junior and senior high school, Greg grew into a large, bearlike kid. His football body was good news for everyone. With a school as small as ours, every boy who wanted could play on the football team. Greg, however, became an especially valued player, not for his speed or coordination, but for his sheer size. With him and Greg White, the other hulk on the team, there was no doubt when our boys ran onto the field that the Liberty Lancers meant business. The general ability of the team was, in fact, mediocre, but our first impression was good, thanks to the Gregs. That fact, at least in the first quarter, gave our team an edge.

Greg White even went on to have a brief career in the pros. Not only was he was big, he was very coordinated and very fast. He played with the Oakland Raiders for a brief amount of time, and that turned him instantly into a Fairfield hero. It was one thing to have a local athlete play well on a college team, but it was unbelievable to have one of our own play on a genuine NFL team. Greg White was, and still is, the town hero because of his tenure in the world of professional sports. Now he farms his family's place and is married to a woman everyone seems to like, a childhood friend of Rick Loeffler's wife, who came from Montana.

Greg Hahner never had the professional aspirations of Greg White, although they both cut an impressive figure on the football field. In a small community, sports are enormously important.

They knit the community together, and everyone, regardless of age, treats the high-school athletes with a degree of parental doting and pride. Everyone was fond of the Gregs, as if we could each take some personal pride in these boys who looked as if they'd been bred for the football field.

Their size and affection for football were about where the similarities ended. Greg White was boisterous and popular, and had a steady stream of girlfriends. Greg Hahner remained quiet, an average student, and to my knowledge, never once worked up the courage to ask a girl to one of the school dances.

He did go, though. Everybody did. The dances happened on irregular occasions in the high-school multipurpose room. Most of the students never danced, regardless of how long they had practiced at home in front of the mirror. Rather, they hung around on the sidelines while the suave junior and senior boys like Jimmy Larsen and Monte Phillips danced their way through select members of the cheerleading squad and the drill team. One would think it was boring for the nondancers, but there was so little to do socially in our community that it was out of the question to miss a dance, even if the event was only an evening of watching the popular kids from the upper grades bop and swing around each other, or, as in the slow dances, cling to each other like barnacles, swaying ever so slightly in time to the music. The lights were turned low, the stereo was pumping out one Credence Clearwater hit after another, the chaperones were all smoking in the adjoining teachers' room, and Jimmy and Karen were barnacled on the dance floor. Everyone else stood on the sidelines watching as if it were some sort of sporting event. This was our idea of fun. Even those enormously shy kids like Greg Hahner couldn't possibly miss it.

That is how I remember Greg as a child. I remember him sitting on the bus in the back seat, reciting Bible verses on Confirmation day, hulk-running onto the football field in the fall, and looming on the edges of the dances in the darkened multipurpose room. Then he went to community college, then Jimmy died, and then Greg took over their farm. All the latter information came to me via my mother's weekly letters, which for the past twenty-five years have followed me around the world, keeping me posted on the neighbors' activities.

It was not until years after high school, when I brought the South American boyfriend home to the farm for Christmas, that I had my first real conversations with Greg. By then, many things had happened, and even though my sisters and I barely knew him, Greg Hahner had become a part of our family. We had all moved away, coming back frequently, but not regularly. We knew about Greg's place in our father's life, though, thanks to Mom's letters.

For the first couple of years after Jimmy's death, Greg stayed pretty much to himself. His Uncle Dick, Claudia's husband, and the man who owned the Appaloosa stallion Chief Qualchan, advised him about planting, fine-tuned Greg's welding skills, and suggested when he should sell his lentils. Dona, Greg's mother, told him what she knew, which was considerable, pulling herself up into the strongest person she had ever been, in the name of keeping her family together. She busied herself with Jill, her youngest child, who was still in high school, and also with learning a new way to help her oldest son be their family farmer. Together they propped each other up after the death of Jimmy, as Greg quietly and steadily farmed. Greg's younger brother, Bryan, was in pharmacy school, and another sister, Ronda, was getting a dietetics degree. Both lived in the dorms at the state university a few hours south.

Dick Dennie shocked everybody when he quit farming in his late forties to go to college to become a psychiatrist. The Cayuse Kings and Queens were long gone, his children were mostly out of the house, Claudia had taken a job at the Crescent in Spokane, registering brides' gift wishes in the crystal department, and Dick

decided to become a psychiatrist. He could have said he wanted
to be an astronaut and surprised no one any more. And with that
decision, the Dennies moved to Spokane.

By then, Greg was a few years into running the Hahner farm
and probably wasn't in desperate need of help, even though he
took to asking my dad for it regularly. It started when he offered to
haul away my father's garbage and then would sneak in a question
or two about disking. Dad had sold his pickup a bit prematurely,
finding out only months after about the new recycling center up in
the valley. Greg picked up Dad's recyclables when he took his own
up. Dad offered him a cup of coffee, a beer, or a piece of the pie
that Mom had left on the counter. Probably every week or two
Greg would stop by for something.

After a few more years Dona married a man named Wayne,
and together they built a house in Fairfield. Greg's sisters and
brother had all finished college, and two out of three were married.
Greg stayed in the family house farming, paying off his siblings'
college tuition bills, and watching television at night. He had a
satellite dish and a subscription to *People* magazine, which he
traded monthly with my parents for their used copies of *Time*. As
near as I know, he didn't have many dates with girls. He was too shy
to speak differently to any of the women he had known as a child.
If there was any way to meet someone new, it was a mystery to him.

By then, my Dad had retired from his second career, assessing
tax forms, and was now home all the time. Maybe it is surprising
that it took years for Dad and Greg to start trading magazines, but
time is a different thing in the country. Redefining how one will
behave with another doesn't happen very quickly, especially with
someone with whom you already have a defined relationship. Greg
was Jimmy's son, and Dad was Jimmy's friend. It took a while to
dissolve that particular generation gap.

Once the ball got rolling it made perfect sense. Greg began
coming over even when he didn't need to take the recycling. He

came to borrow something or help Dad on a project. Then he just
got so he would stop in without any reason at all. During the
winter they watched TV shows brought in on Dad's satellite dish,
drank coffee, and discussed what wheat was selling for. Probably
three or four times a week Greg dropped in, his big bulk hovering
in the doorway as he pulled his muddy boots off on the porch. If
there was no coffee made, Greg put it on. Mom took to making
pies more often. Sometimes he stayed for dinner.

At least once a week they went over to the Harvester in Spangle.
The Harvester could be a restaurant supply store's showpiece.
There is nothing there that can't be bought in bulk: the salt and
pepper shakers, the orange/brown coffee mugs, the simulated-
leather menu-holders, and the colorful, laminated cards that rest
on each table listing the bar's specialty drinks. Hamburgers,
chicken-fried steak, chef's salad, and a dieter's plate with cottage
cheese and a hamburger patty make up the menu. Local women
wait on tables. Most graduated from the high school, moved
away, and then returned after divorcing the husbands they found
in the outside world. It is surprising how similar their current
hairdos are to the ones they sported in high school. Their sons
attend the local high school and drive truck for farmers like Greg
in the summer.

Everyone at the Harvester got quite accustomed to the sight
of my dad, with his cane and John Deere hat, walking in next to
Greg, the bear-man with stooped shoulders. The two ordered the
special, drank several cups of coffee, and moved over in their
booth for other farmers to join them. There were always lots of
jokes and maybe a game of liar's dice to determine who had to
pay. How long the conversations lasted depended on the season.
In the winter those men could be there for half a day.

Sometime between high school and this adult part of his
life, Greg became a sociable person. No one ever described him as
talkative, but he kept up. Farmers, by nature, are reticent, precise

in their conversation, and not excessively eloquent or sentimental. They can sit together for hours—the Harvester a perfect
locale—in silence or discussing at length the price of crops, the
high-school sports teams, or news about other farmers or pieces
of land. Generally they don't tell long stories and they don't
deeply analyze situations. Their senses of humor don't tend
toward the cornball, which is not to say, however, that they don't
enjoy a good laugh.

Greg enjoyed a good joke. You could tell because his giant
laugh took over his entire body. Deep and loud, his laugh echoed
throughout a room. I found it difficult not to laugh with him
when he got going, even if I didn't understand what was so funny.
That made him very likable.

To my knowledge Greg did not go to movies with his peers,
did not go on dates, and did not spend weekend nights in the bar
at the Steak House, getting drunk and yelling gross things that were
funny only to the other drunks in the room. His social life consisted
of dining with his family, a large extended group that stretched out
over the school district, hanging out at the Grain Growers or at the
Harvester, and visiting with my father. Apparently all of this was
quite sufficient for this man in his early thirties.

Having his own farm, Greg could share anecdotes about
machinery breakdowns, stories about field management, and
tales of his own relationship to grain prices. He took on the
mannerisms of every farmer in the area. There is the slow, rolling
walk, never rushed, whether it is between the house and the barn,
or the church parking lot and the steps. There is the way they rest
their hands in their overall pockets during a conversation, with
their thumbs still sticking out, as if at any moment they will be
back to work. There is a dry, salty sense of humor punctuated by
guffaws and shoulders that shake with laughter. Greg's body
became that of a farmer: thick, like a wintering animal, and
strong, like an ox. Endless hours out in the eastern Washington

sun had left him with parched, weathered skin, the kind that cracks at the cuticles and elbows and makes a sandpaper sound if you rub it. His hands were always slightly tinged with the dark gray of machinery grease. Even on Sundays there was a black ring around each cuticle.

None of us girls were jealous of Dad's affection for Greg, even though it didn't take Dick the Psychiatrist to see that Dad preferred his company to ours. We would come home for visits only to have Dad stand up and amble out the door when it was time to go to the Harvester, even if he had just been there the day before and we had just arrived. If Greg drove up, his garbage cans in the back of the pickup, Dad pounded on the window and motioned him in. It seemed important to Dad that Greg know he was welcome, even during the infrequent intrusions of the daughters. From our perspective we were just glad. Dad seemed the happiest he had been since that dreadful time when he had decided to quit farming. It wasn't that we felt guilty for being girls, but all of us at one time or another had wished desperately, for his sake, that one of the others had been a boy. Greg took up some of the slack that had always been dangling around the farm. We all felt better.

One Christmas we all came home. By then I was in my early thirties. Whereas my sisters brought their husbands, I brought with me Carlos, the Ecuadoran. Greg was especially keen to meet Carlos. He was more interested in foreigners now since his sister Jill was living in Europe, where she was married to an Iranian. Even though she hoped to bring him to the United States soon, Greg had been thinking about a trip to Europe. "Why not? I've been to Disneyland. That's like a foreign country, isn't it?" he said. Carlos's limited English didn't allow him to get the joke.

Carlos had never seen snow before, which spurred Greg to bring his snowmobile over every day during the visit. He taught Carlos how to jump ditches, to skirt the draws, but not about chasing coyotes, a local snowmobiler sport Greg didn't subscribe to.

For someone who seemed slow to make friends, Greg formed an instant bond with Carlos, much to our surprise. Although this was his first trip to the United States, Carlos had been all over South America and Europe. He was sophisticated and worldly. A main avenue in Quito was named after his great-grandfather, and the head of the National Bank played tennis with Carlos's father. It was surprising that his English wasn't better, considering all those years in private schools, but some people just don't have the knack.

In front of him on the snowmobile was Greg. He had been to Disneyland, gone through twelve years of public school in rural America, and spent one and a half years at a community college in Wenatchee, Washington. He had to dig out an atlas after meeting Carlos to figure out where on Earth Ecuador was.

The two of them went on long snowmobile rides. When they returned, we saw them from the kitchen table, standing for another hour at the side of the machine. Carlos would be waving his arms, clearly recapping some highlight that had just happened on the snowmobile. It had been too noisy to discuss it at the time, so now the two reviewed the whole ride, remembering the best ditches and the highest hills. Greg stood there with his hands in his pockets, his shoulders bouncing up and down in laughter. How many of these lengthy Carlos stories were actually in an English that was understandable and coherent was a question. It didn't seem to matter.

On Christmas Day, Greg ate turkey with Dona and Wayne, but on New Year's Eve he came to our house to drink hot buttered rums and play the Pictionary game someone had brought from California. Carlos and Greg were a team. Greg's huge laugh shook the room when Carlos managed a correct word in English or Spanish. Dad just watched it all, smiling a bit. He draws like he walks, kind of slow and limping. Outside of liar's dice and a little gin rummy, he avoids games.

After New Year's, Carlos returned to Quito. Ours was a fleeting romance born out of my pursuit of a master's degree thesis dependent on research done in Ecuador's Amazon jungle. We daughters returned to our cities, Mom went back to her class, and Greg and Dad returned to the Harvester.

In June Greg got the flu. It lingered too long, and his few trips to the doctor turned up nothing. Then one morning his hired man drove up to the house and found Greg unconscious on the floor of the kitchen. At the hospital and after many tests, the doctor told Greg, who had come to, that he had a brain tumor. It all happened very quickly.

Before surgery Greg announced to everyone in the room that if he were going to be like the Mona Zehm in his vague child-hood memories, struggling just to make it down the church aisle, he did not want to live. He was ten years younger than Mona had been when her brain had erupted. Dad stood at the foot of Greg's hospital bed. Greg remembered that my father had had brain problems also, although of a different sort. "Well, Ralph, we got another thing in common besides the draw by the creek and recycling day."

Dad drove to the hospital every day before and after Greg's surgery. When not at the hospital, he sat at the kitchen table and looked out the window, as if Greg would drive up with his used *People* magazine on the seat next to him. Dad just waited. He didn't watch TV. He didn't work in the rose beds. He didn't eat any pie. Sometimes he did stop at the Harvester, where the waitresses looked at him with concern. One day they gave him a card to take to Greg. Everyone from the restaurant had signed it, even several regular customers.

After the surgery the doctor said they couldn't get it all, but they could give Greg chemotherapy. This would require weeks of frequent trips to Spokane. Greg was unconscious as the doctor explained this to Dona and my dad, who said, "Well, Greg's given me more than one trip to the doctor. I guess I can return the favor." A few days later Greg opened his eyes and said, "Hello," but that was about it for him. He went into a coma after that "hello" and died two weeks later. After what he had said about Mona, no one was surprised.

* * * * *

An organist from Spokane played at Greg's funeral. Everybody in town was pretty choked up about this death, and nobody who knew Greg thought that she could play the organ and get through the service at the same time. The funeral was at our Lutheran church. Some cousins carried his casket, although on the program Greg's mother had named Dad as an honorary pallbearer. He walked behind the casket. Mom thought Dad wouldn't make it through the funeral, but he did, in that way he seems to make it through everything, in silence and with dignity. He did express a bit of embarrassment to Dona, to be so old and to follow that casket down the church aisle behind the thirty-year-olds. She told him quickly through her pain that Dad would have been first on Greg's list, if he ever would have imagined he had to make one so early in his life.

One year later Jill, Greg's youngest sister with the Iranian husband, discovered she had the very same kind of brain tumor that Greg had. She was twenty-six years old. She had her surgery in Seattle and lived for two more years before it killed her. When she was dying she said to Dona, "No funeral. We've had enough."

Jimmy, Greg, and Jill have side-by-side graves in the Fairfield Cemetery. Their markers are identical except for the names and dates. Each has one stalk of wheat leaning over, as if in a breeze, bending toward the Tekoa Mountains.

My father has always had an almost obsessive way of throwing things out. He has given new meaning to the word *tidy*. For the most part, anything that we hadn't used within the past year or two got tossed into the back of the pickup and hauled to the dumpsite down by the creek. We had the neatest, most rubbish-free farm in the district. There wasn't so much as a rusty pail left behind a shed to gather rainwater in the spring. Dad folded all the usable burlap seed bags and stuffed them into the empty fertilizer barrels that lined the tack room of the horse barn. He stacked the laths neatly in the big barn, up against the north wall. If we took one out to use in herding the cattle, Dad expected us to put it right back. We swept bent nails off the shed floor, seed packets went into the burn barrel, and Dad told us to toss dog combs missing more than four teeth. Every time a clumsy steer stepped on a cat dish, we replaced it with another hubcap. Truck hubcaps were the best, big enough to feed the unknown number of cats that always showed up at mealtimes, many nameless and of unknown origin.

From the hill near Uncle Detlef's, our farm looked like a photograph from a social studies textbook, the kind that city kids read to learn about the country. "The golden wheat fields that surround this farm help provide nourishment for the entire world" would be the caption on the photo.

It is the quintessential farm. The barn is classic red with white trim and giant sliding doors on every side. The two-story house has a huge porch on one end. Willow trees, lilac bushes, and an emerald-green lawn surround it, thanks to that water witch who found us a good well that produces even at the end of summer. On the east end is a flower bed that my parents tend religiously. The outbuildings aesthetically dot the spaces between the house and the barn. The fields frame the farm, in undulating waves of golden wheat or green seedlings or brown dirt, depending on the season. The whole place was and still is in perfect order. There is not so much as a gum wrapper or a bent nail in sight.

When I was a child there were garbage barrels everywhere—not in plain sight, of course, but discreetly tucked into corners of tack rooms or behind grain-bin doors. This was long before the days of recycling, and everything got tossed with great abandon into the barrels. With regularity Dad loaded the full cans into the back of the pickup and took them to the dump.

The dump was in a pristine location up the road from our Hangman Creek piece. I don't know if the county had officially named it a dump, or if the general accumulation of people's castoffs over the years, a waterfall of garbage cascading over the cliff, had, by default, designated the area as such.

We loved to go to the dump with Dad as long as he would let us jump from the pickup cab before he backed up to the edge of the cliff. We were certain, with each trip, that the brakes would fail and the pickup and all of its contents, be it garbage or our father, would go falling down the canyon wall. It was thrilling to watch Dad edge back little by little toward doom, only to brake at the last possible moment, a wave of relief washing over us with the knowledge, once again, that we wouldn't be orphans. While Dad tipped the garbage cans over, one at a time, letting the useless remnants of our lives tumble through the air and disappear into the canyon below, we picked through the useless remnants of other people's lives, people who were not as conscientious about getting their trash over the cliffside as we were, or perhaps weren't as brave as our father in their willingness to back their vehicles right up to the edge of the cliff.

When I think about it now, I think that my father, in many ways throughout his life, was often backed up to the edge of a cliff.

We rarely found anything of value on those trips to the dump. I always knew the good stuff was resting at the bottom of the canyon in some place far from any road. I don't even know who owned that part of the creek or how we would have found that pile of trash. I found a rope once, as thick as Dad's arm, which we took home and tied as a swing to the cottonwood tree. Another time Cheryl found a metal All detergent pail that was completely fine except for one bullet hole at the very top and a loose handle. We didn't need another pail, but Dad let her take it home anyway. For weeks afterward she would water her horse with it, as if Lady Anne were incapable of ambling over to the watering trough like the rest of the horses. With black paint she wrote Lady Anne's name in giant letters across the side so that no other horse would make a mistake and drink from that bucket.

My father instilled in all of his daughters a love for throwing things away. To this day there is nothing I love better than to use the last drop of shampoo and hear the empty bottle fall into the wastebasket. Peeling the last square from a roll of paper towel gives me an enormous amount of pleasure. I will never buy those giant boxes of laundry detergent or economy-size bubble-bath jugs at warehouse stores. It takes too long to empty them. In this regard I am an environmentalist's nightmare.

As children, we scraped the very last drop of horse-hoof salve from the can with great glee, rubbing it over our ponies' cracking hooves with care, and then immediately tossed the can.

We lathered up those saddles in the summer heat, using an extra smear of saddle soap just so we could empty the tin and listen to the echo as it hit the bottom of the garbage barrel. We fed the dogs extra food, just to have the satisfaction of shaking those last nuggets of Purina Dog Chow out of the giant paper bag, crumbling it up and stuffing it into the burn barrel. We loved gathering old clothes and toys and giving them to the Goodwill. It had nothing to do with a desire to share with those less fortunate than ourselves. We just loved the absence of clutter, the large spaces left by departed possessions, a notion of having less responsibility with fewer possessions.

I would not describe Mom as a pack rat, although as the sentimentalist of the family, she was on constant patrol, waiting for someone to trash something that had lasting monetary value or, more importantly, sentimental value. You could divide the category of sentimental value into three subcategories: things belonging to people who were now dead; things that belonged to an adult when they were a child; and things that now belonged to a child who, when turned adult, would possibly find them a treasure. This last category was a bit difficult since it essentially designated everything we girls possessed as material for Mom's sentimental journey, assuming we lived to adulthood. But then, if we didn't, our belongings would immediately be funneled into the dead-person category. The categories were so broad that Dad ignored them completely and tossed freely.

This has been known to drive Mom nuts. Dad admits throwing out the Melmac picnic dishes we used on that endless car trip we took to North Dakota to visit Mom's side of the family. To cut corners we picnicked most days in city parks along the way. When Mom discovered that Dad had tossed them, he explained that the dishes were scratched and faded.

Mom felt bad. I think she was thinking about those little Vienna sausage sandwiches we made and how they looked on those purple plates. Also the way, after we finished our sandwiches, we rinsed the dishes from whatever park spigot we could find. While the dishes dried in the sun, sitting on top of our car blanket, the whole family would stretch out on the park grass. Maybe those dishes helped her remember how the sun would feel on our front sides, hot and penetrating, long before the days of ozone layer and UV ray fears. Or the way the grass would scratch our backsides as we lay stretched out on that Idaho, or Montana, or North Dakota small-town city park. It was such a relief for us to peel out of our car, a Ford Mercury, I believe. We lay there squinting at the sun, knowing that the clock was ticking. Soon we would have to get back into the car, roll down all of its windows, and drive on for another five or six hours before we could start to look for a motel. Mom would break up the ride by cracking open two Cokes in the midafternoon, steadying the purple Melmac cups on the open glove-box door as she divided the pop between the cups.

Dad also admitted throwing away Grandma's butter churn. One of the wooden paddles was cracked. He tossed the wicker rocking chair that had sat lopsided in the old coal room for decades. After it was long gone, we found out it had been the rocking chair of Great-grandma Keller, who had died at age 103. She died at home, maybe even in that chair, for all we know. Dad threw away the giant crock in which Grandma Hein made bread-and-butter pickles, and he threw away the old mule harnesses that the grandpas used before tractors pulled the swathers. He swears he didn't throw away the old brown telephone, the first receptor for the party line, but does admit he tossed the portable diving board when the springs rusted out.

* * * * *

Did Chief Qualchan and the other Indians know how to swim? I asked my father that question on the day we all were down at Hangman Creek putting up the picnic table by the monument. Nobody there seemed to know. I liked the idea of the chief and me jumping into the same water hole. This was assuming not only that he could swim, but that before Colonel Wright took up his short residency at what was to become our Creek Piece, Qualchan had had the opportunity to float in these waters.

That's where my father learned to swim, or not swim, as he would say, because in truth, he has never been very good at it, thanks to the sink-or-swim lessons of Grandpa Hein. All Dad learned was how to thrash around enough to get to the edge of the bank, however ungracefully.

When we were kids, Dad found a portable diving board on sale at the Kress store in Spokane. It was small and wooden with rollers underneath. The board, attached to a large spring, wasn't very big but could easily support the weight of a child. Dad would take us down to the creek, prop the board on the bank, and let us jump off into the deepest part of the water. Unlike Dad, whose father threw him in whether he liked it or not, we begged for the diving board, having learned early on that with vigorous dog paddling we could get back to the muddy bank and another turn at jumping off.

Our whole family spent many an afternoon down at the Creek Piece, usually on Sundays. We drove down in the afternoons after lunch. First we counted the cows. They spent the spring and summer there. Dad had fenced in our property, a section that covered several acres. We tracked down the herd, with each family member counting until we agreed on the numbers. They were almost always all there, placidly grazing, except for the sad time after a lightning storm when we arrived at the creek to find two cattle dead under a fallen tree. On another occasion we arrived to find one steer missing and nothing but a brownish pool of drying, bloody ground to tell us someone had shot the unfortunate animal and carted him off for stolen winter steaks.

Usually all the cows were fine, and once we had counted we could go swimming. Dad set up the diving board and watched us while we plunged and wallowed in the mud. Mom set out a picnic, finding a spot that was absent of cow pies and thistles. If she finished setting up before we finished jumping, she took a walk to look for arrowheads. She always said she used to find them, but I never saw any. Then we all ate the picnic together, lazed in the sun a bit, and maybe we girls took a few more jumps into the pool before we loaded up for home.

On the day we Cayuse Kings and Queens rode our horses down to the creek to put up the picnic table, Dad drove the diving board down in the back of the pickup. All the club members knew to bring a towel in their saddlebags and to wear a swimsuit under their jeans and cowboy shirts. While the dads hammered and sawed, we equestrians raked pine needles and then jumped and squealed in the water, our horses grazing on the flat above the creek, their bridles replaced by halters so they could eat more easily.

There were no turtles, but we think the deep part must have been the same place the children from Rattlers Run School took to playing with them, not to mention the same place Fat Louie took his last dive.

Things have changed some in our neighborhood since that picnic-table day. Of the ten farms, two are no longer inhabited. The Creek Piece looks pretty much the same. The picnic table is gone, although the monument is still there, same as it ever was, erected by the Waverly Historical Society in 1938.

The Dennie house has been empty since they moved away when Dick went to college. The Dennie kids are spread up and down the West Coast, and I almost never see or hear of them. While the fields are still farmed, the house has been let go to become a haven of peeling linoleum and mountains of mouse turds.

A family nobody knows lives in the Hahners' house. They rent it from Dona and work somewhere else. Dona leases the

farmland to a farmer on the other side of the creek. She lives in
Fairfield with her second husband, while Bryan and Ronda live in
Spokane with their families.

The Brewer place is empty as well, and each year that house
sags closer to the ground. Ed died of his cancer right after he and
Harriett moved to Spokane. Brenda married a Mormon and lives
in Phoenix. At first she didn't come to visit very often because she
had so many children they couldn't all fit into one car. Now that
some have gone off to college or are on their missions, she gets up
more often. Mary lives in Idaho with a husband I've never met.

Steve Martinson lives on his family place. He doesn't farm,
though, but prefers to work at the high school down the road.
When Barbara died of lupus, Joel moved down to Waverly. His
mother lived down there and could help him with the kids.
Belinda, the oldest, was ten at the time, and Stevie, the youngest,
was only six. Not only do the three Martinson kids now have their
own children, but Belinda herself is a grandmother, a realization
that puts a time warp on everything for me.

Dick Flaig, Marsha's former boyfriend, has had a wife or
two, and I believe a career as an architect. Darryl, his younger
brother, lives on their family place up Cahill Road, farming their
piece, not to mention ours and several others in the area. Farms
must get bigger to pay the bills these days, and people have their
eyes on their neighbors' property, although not necessarily the
farmhouses that come with it.

Leonard Zehm married a nice woman named Beth after
Mona died, and they moved into a manufactured house near his
dad up on the old Thams place. His house was starting to lean
also, but Devin, Darryl Flaig's oldest son, is living there now with
his family. Devin shored up the house, and his wife planted a row
of poplar trees to grow into a windbreak. Between their recently
born first baby and the farm improvements, the old Zehm place
is looking quite revived. Dwight, Leonard's oldest, lives in Alaska,

and Barry, Leonard's third son, lives in Arizona. The rest of his children, with their children, are scattered between.

The neighborhood got a little national attention when one of Darryl's fields became a set for the Hollywood movie *Toys* with Robin Williams. An article in *The New York Times* arts section said the script called for a "surrealistic landscape," and Lord only knows how Barry Levinson, the director, found us. They needed wheat fields that stretched out forever, rolling across a vista that buildings did not interrupt. Just that whisper of Mica Peak in the background assured us that this was our part of the Palouse up on the big screen. The Hollywood company had to film in the spring to capture the rich lime-green that the wheat fields achieve for a thread of time on their route to becoming the golden waves of grain of textbook and song fame. When filming took longer than they expected, the company spray-painted the wheat to keep it the right color. I guess they paid Darryl well enough for the venture to sacrifice that particular field to a can of paint.

* * * * *

The idea of sacrificing fields has taken on a new meaning to us by now with all the news from and of Hanford. In 1986, Department of Energy officials reluctantly released 19,000 pages describing the secret operations of Hanford during the forties and fifties. After incidents such as the Three-Mile Island nuclear disaster, as well as stories in both the Seattle and Spokane papers reporting radioactive leaks from Hanford, there was public pressure for more information. The 19,000 pages revealed details about the releases of radioactive materials into the Columbia River and over more than 75,000 square miles of land.

In the business of making plutonium, it was routine to release a certain amount of radiation into the air. Even though evidence exists that the risks of exposure to radioactivity were known at the time, the business at hand was to make a nuclear bomb as quickly as possible. What the papers revealed was that Hanford didn't want to take the time to cool the fuel long enough to significantly reduce the risk of exposure. Iodine-131 has a half life of eight days. If the fuel had been cooled for ninety to one hundred days before being released into the atmosphere, the risk would have been much less than it was with the usual thirty- to fifty-day wait. Large amounts of waste products, such as iodine-131, plutonium, strontium, cerium, and ruthenium, were spewed out over the land and into the Columbia River. These emissions, which occurred between 1944 and 1972, with the bulk of them in the late forties and fifties, were hidden from the public.

In 1985, *The Spokesman-Review* published its first article about "downwinders," a term that refers to anyone who lived in the areas where the winds carried radiation. The headline read "Downwinders—Living With Fear." My mom cut it out and sent copies to her daughters.

There are more questions than answers in the case of Hanford and health. Scientists argue about whether short, intense exposure to radiation, as in the cases of Chernobyl and Nagasaki, is less likely to cause cancer than numerous smaller exposures over a longer period of time, as in the case of Hanford. The insufficient weather information for the twenty-five years the releases were occurring promises inconclusive reports. Comparison studies between places such as Chernobyl and Hanford are made more complicated by the differing diets and lifestyles of their residents.

In spite of these factors, there are some knowns, some conclusions, and a fair amount of inklings. Iodine-131 is released as a gas into the air, landing silently on fields and infecting both the milk people drink and the vegetables they eat. Scientists think there is a direct correlation between thyroid malfunctions, cancer, and iodine-131. Radioactive particles of plutonium, strontium, cerium, and ruthenium attached themselves to rust or dust as it left the nuclear plant. Plutonium exposure can cause ill-health effects such as bone, liver, and lung cancer; leukemia; and chromosome aberrations. Strontium can cause leukemia and bone cancer and weaken an immune system. Autoimmune diseases include multiple sclerosis and lupus. Cerium can cause leukemia and bone, liver, and nasal cavity cancers. Ruthenium can cause unspecified cancers and skin burns.

My mother was the one in our family to find the Hanford Health Information Network, a government office funded by Congress in 1990 to provide information regarding the plant and its history. This information comes to us slowly, periodically dropped on us as papers are declassified and studies completed. From this organization, as well as others, we have learned many things, including:

- Radioactive waste from Hanford was traced to places all over eastern Washington, and into Canada, Idaho, and Montana.

- In 1945, Hanford's plutonium plants released 555,000 curies of iodine-131. A curie is a measure of radioactive material. One curie is 37 billion atoms undergoing decay each second.

- In 1949, the "Green Run" occurred, a planned release of approximately 8,000 curies of radioactive iodine-131.

- In 1951, special iodine filters began to fail, and before they could fix them, 27,000 curies of iodine-131 were mistakenly released over eastern Washington during the height of the growing season.

- From late 1944 through 1951, there were large releases of particles containing plutonium, strontium, and cerium.

- From 1952 to 1954, there were large releases of particles containing ruthenium.

• Most of these releases, because of size and weight, landed around the Hanford plant, but particles were detected as far away as Spokane, Mount Rainier, and even Montana.

• In 1956, Hanford officials considered issuing restrictions on drinking untreated Columbia River water, but decided against it since the restrictions were "not essential" and "public relations might suffer from such restrictions."

• In 1973, the *Seattle Post-Intelligencer* reported that 100 billion gallons of low-level radioactive liquid waste were discharged into the ground during thirty years of Hanford operations.

One thing led to another, and before we knew it, lawyers were flying Mom to Seattle to testify on behalf of the residents of Zone Four, which is what our area was informally named by the lawyers to identify our proximity to Hanford. She had signed our family up in the lawsuit against the companies that had the government contracts at Hanford in the forties and fifties.

Many people in Fairfield prefer to think of this activism as left-wing nonsense. My parents have always been liberal Democrats in a sea of conservative Republicans. Putting the left-wing stuff aside, it's not so surprising that people could be upset. If you're sick, you want answers, but if you're not, you want reassurances.

It's hard to look out over those fields, a view that has remained virtually unchanged for the past hundred years, and consider the hills as poisoned. Those are our hills, and it angered people that Mom seemed to blame the land for the illnesses. She doesn't claim to be sure—no one does—but one can't read the government papers, papers they were reluctant to show us, without a fair certainty that the questions are worth asking. If it turns out that we all got poisoned for the war effort, we can know that it is more than just enlisted people who were sacrificed. Perhaps we residents of

Zones One through Five will one day be awarded Purple Hearts for our heroic, albeit unknown, contributions during both World War II and the Cold War. It seems only fair to me.

Who would want to believe for one second that Hanford could be blamed for many things, from my father's inability to do the jitterbug to Jimmy and his children lying dead in the Fairfield cemetery, hideous bookend events that hold in the suffering and grief that our entire neighborhood has experienced over four decades? Who would want to imagine that someone poisoned our hills, sometimes knowingly? Who would want to believe that in the name of stopping the Evil Russian Empire from swallowing us up, companies, under contract with our own government, would poison all of us, without at least giving us a chance to get out of town?

And besides, even if there is a known connection between radiation and some of the cancers that occurred in our area, there isn't to all, so how do we explain them? Well, it could be that those are just cancers that would have come our way anyway, or maybe we just don't know enough yet. Extensive studies aren't completed on the relationship between weakened immune systems and cancers in general. There also exists the possibility that people who have died of other illnesses besides cancer may not have contracted their diseases but for an irradiated immune system. Everybody says that it sure seems we've got a lot of people sick with multiple sclerosis in our area, including Susan Felgenhauer, who, with her sister Sally, was in the Cayuse Kings and Queens with us. Susan and Sally were the first two off the bus when it took the winter route. Susan still lives on her family farm, now with her husband, who farms for Susan's dad. There is a rumor that our area has the highest rate of multiple sclerosis in the United States and the second-highest rate in the world. Joel Martinson told me this on the same day he spoke of the astronomically high rate of lupus we also have.

And then there was the lesson of Mount St. Helens. With that mountain's explosion we learned a lot about vulnerability and how far the wind can carry things.

Everyone in the Northwest remembers where they were when they found out that John F. Kennedy was shot and when Mount St. Helens erupted. We had been hearing for so many months that either it or Mount Baker was going to explode that we had given up thinking about it. I lived in Seattle when a neighbor, on that May morning in 1980, pointed out the vertical white cloud that stuck up from the south like a stalagmite escaped from a cave. Seattle is about a hundred miles north of St. Helens. All we saw was the white plume. The rest I experienced via the gruesome images on television and speaking on the telephone with my relatives in other parts of the state.

My parents were at church in Fairfield when the pastor made the announcement. He kept the prayer short after the offering, because, thanks to a news report, he knew an ash cloud of immeasurable size was lumbering toward them. Fairfield is about 350 miles from St. Helens, but the winds were blowing in its direction. On their way home from church, my parents listened to instructions on the radio about what to do. Dad put all the cats he could find in the shed with the doors shut tight and the food and water bowls full. Mom closed all the windows in the buildings, the cars, and the pickup, and for only this once, brought Cleo, their Great Dane, inside the house.

Dad was standing in the driveway when he yelled to my mother to come look. At one in the afternoon, there it was, a giant cloud wall of tumbling dark gray, rolling in from the southwest. For hundreds of miles it had ridden the winds blowing toward Idaho, leaving a blanket of ash in its wake.

By two in the afternoon, the sky was completely dark, a starless midnight to last for who knew how long. My parents, holed up inside the house with the dog, described an eerie silence as the ash drifted through the air outside. They hadn't had any previous experience with this sort of thing and could only sit in front of their television listening to the news as reporters described what was happening all over the state. They went to bed early and awoke in the morning in an obscure light that barely allowed them to make out the outline of the barn through the haze.

The last of the ash came to rest in parts of northern Montana and Saskatchewan, Canada, many hundreds of miles away from the source. For weeks after the eruption, my parents shoveled ash into a giant pile in their driveway, a pile that towered over them and took nine dump-truck loads to banish from our farm. That summer no one went outside without wearing a white surgical mask in the hope that it might help somehow keep our lungs cleaner.

If you thought about Mount St. Helens, 350 miles away, it wasn't too difficult to believe that the wind could easily carry radioactive air garbage the 100 miles from Hanford and invisibly layer it over our fields, buildings, and selves. The newspaper published a map. The map showed a distorted, wobbly dartboard illustrating the areas where they found radioactive iodine-131 on the vegetation. The bull's-eye sits exactly over the nuclear site. Its darkest color indicates the highest levels of radiation. An outer, slightly lighter oval lists to the northeast, thanks to the prevailing winds of the day. It expands past the towns of Washtucna and Ritzville. Yet another larger, even lighter oval finishes the spread up to the northeastern part of Washington and down into a giant oblong shape covering the middle two-thirds of the state of Oregon. I grew up in the middle oval, the second-darkest color, the second most intense concentration of radioactive iodine-131, and what is called Zone Four in the lawsuit. The map implied that the closer to Hanford, the more radiation one received which, in fact, was not necessarily true, thanks to the magic of prevailing winds.

I don't know what was strongest in my mother's mind when she filled out those papers to add our family to the lawsuit: my father's cancer, Mona's death, Greg's tumor, the other neighbors' illnesses, or the screwed-up thyroid glands that she and my three sisters now deal with on a daily basis.

My mother lived on our farm for more than fifty years. She is the kind of person who writes letters to the editor, as well as to anyone else she feels she should inform about a topic. Years ago she and my father had a dispute with one of the absentee landlords they farmed for. In addition to our own land, Dad farmed the nearby fields owned by other people. Sometimes the union of landlords under the tutelage of their farmer can become complicated. With the fluctuation of grain prices and seed rates and other factors, the landlords must trust in their farmer to make decisions about their land while tending the needs of all the other landlords. The landlord in the dispute was the wealthy local lawyer, a man who claimed farm subsidies that my parents felt were more fittingly deserved by their other landlords. Earlier in the year, the lawyer had made an agreement with my parents regarding those subsidies, but then he changed his mind. When he changed his mind my parents took him to court. When the local courts ruled in his favor, Mom took it to the state court, and when they ruled in the lawyer's favor, Mom pushed the case to a special court in Washington, D.C., that specialized in agricultural concerns, writing lengthy accounts to the judges there, who eventually ruled in favor of my parents. They never stood to gain anything from the battle because the profits and losses concerned were those of the landlords. They stood up to the lawyer because that's the way they are. In our family, wealth and power in and of themselves are neither virtues nor intimidating.

Before he died, the lawyer called my father from the hospital to tell him that his lawyer made him do it. We think the lawyer should have stayed a farmer, like his father, who gave him the land. We think his work might have made him go sour.

A woman with my mother's history of righteous conflict resolution had no problem reporting to the people who were organizing the Hanford lawsuit. They flew her to Seattle, and she gave her deposition to a group of lawyers for General Electric, Rockwell International, UNC Inc., Westinghouse Electric, and Atlantic Richfield, the companies being held responsible for those puffs of radioactive waste that landed on us.

At her testimony in September 1991, Mom started her account with our neighbors. The Crabtrees lived about three-quarters of a mile from us. Emma had breast cancer in the fififties, but both she and her huband, Harley, died from Hodgkin's disease. Gordon, their son, is still alive, in spite of his bladder cancer, although he no longer lives in the area. Next to them was Detlef and Mary's farm. Detlef contracted bone cancer in the thirties, long before Hanford could be the culprit. Moving in a northwesterly direction, the next family is the Zehms. Mona died of a brain tumor, and Gary, her son, survived his testicular cancer. Then come the Hahners. Jimmy is dead from colon cancer, and, of course, Greg and Jill are both dead from brain tumors. After the Hahner family is the Martinson family. Barbara died from lupus. Ed Brewer died from pancreatic cancer, and his son, David, from leukemia. To complete the neighborhood rundown was my dad's thyroid cancer, and then a melanoma, neither of which killed him, thanks to surgery. Then there is the rest of the thyroid stuff. Both Ed Brewer and Greg Hahner had thyroid problems before their deaths. Mom doesn't know about any more neighbor thyroid problems, but that's not necessarily a thing that would come up in conversation. All the women in my family, except me, take medicine to regulate that now-infamous gland. Our neighborhood has had so much that we don't necessarily mention the little things. Anything short of cancer, or something equivalent, doesn't get much attention. "This square-mile neighborhood I have just described has only ten families living in it," Mom said. "Of the ten farms, seven have had at least one, if not multiple, cases of cancer in the last thirty years."

It's not that we're certain our neighbors are dead because of Hanford. Maybe it's DDT or bad genes or just plain bad luck. But maybe our little neighborhood is what the Hanford investigators call a "hot spot," a place that by virtue of such factors as wind patterns, precipitation on key days, existence of hills and mountains, or simply being the location where the winds died out, subjected us to unusually high doses of radiation.

We've considered all the possible causes. At Jill's graveside one of her cousins said, "That does it, I'm not eating any more of Aunt Dona's potato salad." Maybe this whole thing is Dona's fault.

* * * * *

Recently I was thinking about Greg. Specifically, I thought about the way he hauled off my dad's garbage, a way of beginning or at least redefining their friendship, long after the days when we dumped our own garbage down by the creek. There still could be a million treasures there—maybe even our old diving board hiding under some huge pile of ancient trash just up the canyon. It's been years since anyone dumped there, but I imagine remnants still cling to that cliffside not too far from both Fat Pond and where Chief Qualchan took his own final dive. He probably did do all those horrible things that Colonel Wright claimed. Why wouldn't he? They were at war, and who can blame him with Colonel Wright and his acolytes roaming the Pacific Northwest, claiming territory right and left? Qualchan was probably just trying to keep his head above water in the only way he knew, swimming through the flood of Colonel Wrights and people like my family, who just kept arriving.

I see a kind of connection between our neighborhood illnesses and Qualchan's murder. Both incidents are about this piece of land, how people should be able to live on it, and about the intrusion of outsiders. Qualchan killed outsiders, and Mom and Dad joined a lawsuit.

And if you'll go with that connection, maybe you'll see that I could be likened to Qualchan's wife, Whiet-alks. As Colonel Wright's men dragged her husband to the hanging tree, she drove her lance into the ground and galloped off on her Appaloosa to tell her story of how it was. Well, it might be stretching it a bit, but this is my book, and here I am, writing this account of how our story has been. To tell you the truth, I have always wanted to compare myself to an Indian princess, especially the beautiful Whiet-alks.

epilogue

Roughly four thousand people are suing the five private contractors that ran Hanford for the government during World War II and the Cold War. Over the years the claims have been consolidated into two major suits. Most of the plaintiffs are from eastern Washington, with a few from northern Idaho and eastern Oregon.

In 1998 and 1999 U.S. District Judge Alan A. McDonald of Yakima severely limited the number of people eligible as claimants when he ruled that in order to qualify, people had to prove they had received a dose of Hanford radiation high enough to cause twice the number of cancers as would occur in the population at large. Most of the plaintiffs in the case did not receive doses that huge; rather, they received numerous smaller doses over a span of more than two decades. Over the years, Judge McDonald has repeatedly indicated that he wants this case quashed.

Lawyers for the plaintiffs appealed Judge McDonald's decision, and in the spring of 2002 the U.S. Appeals Court for the Ninth Circuit overturned his ruling. As expected, the defendants in the case, General Electric, Rockwell International, UNC Inc., Dupont, and Atlantic Richfield, appealed the appeal, slowing the march of this case toward a jury trial, which the plaintiffs remain optimistic about winning. Historically, the party with the most resources, in this instance the defendants, strives to slow the case and dry up the plaintiffs' resources before it can go to trial. It should be noted, though, that because of provisions in the Hanford plant's operating contracts, the federal government reimburses the former contractors for their legal expenses, which so far exceed $60 million. In October 2002 the Ninth Circuit rejected the defendants' rehearing petition and remanded the cases back to the U.S. District Court for Eastern Washington. Now attorneys for the downwinders are busy making plans to resume litigation.

At the present pace, it could be another ten years before the case is resolved. My father will be ninety-two years old in ten years. He was thirty-one when he was diagnosed with cancer.

Reports of radioactive waste leaking from aging storage tanks at Hanford are prevalent today. There are 177 high-level radioactive waste tanks, each holding as much as a million gallons—an estimated total of 54 million gallons of the most dangerous waste on the planet. Of those tanks, 149 are single-shell, and the oldest of them were designed to last ten to twenty years. The oldest tanks were built in 1944 and are still in use today. Sixty-eight of the tanks are known to have leaked more than one million gallons and possibly up to five million gallons of waste into the soil. Stewart Udall, a former congressman and secretary of the interior, has called Hanford "the single most polluted place in the Western world."

Site officials now admit that significant radiological contaminants are seeping into the groundwater. It will not be long until this radioactive waste reaches the Columbia River and is swept downstream; some scientists say that is already starting to happen.

Government efforts to contain the waste have been so mired in bureaucracy that in the year 2000, the attorney general of Washington State, Christine Gregoire, sued the federal government to force it to begin pumping free liquids from the leaking single-shell tanks immediately. The Bush administration's Department of Energy (DOE), in an effort to cut cleanup costs, has proposed reclassifying the highest-level waste as "incidental waste." By law, high-level waste must be vitrified and ultimately removed to off-site disposal locations. "Incidental wastes," however, can be disposed of in shallow land burials at Hanford at a significantly lower cost.

The Natural Resources Defense Council (NRDC) and the Yakama Indian Nation have filed a lawsuit challenging the DOE's attempts to circumvent the law and give itself the power to reclassify the waste. The states of Washington and Idaho have entered appearances in the lawsuit siding with the NRDC and the Yakamas. The judge in this case recently disallowed a motion by the DOE to dismiss the case, which is now heading for trial in U.S. District Court in Idaho.

The Hanford Thyroid Disease Study, a twelve-year study of some 3,400 people born in the affected area, found no link between thyroid disease and Hanford's radiation releases. The study was funded by the Centers for Disease Control and conducted by the Seattle-based Fred Hutchinson Cancer Research Center. The study's final report in June 2002 concluded that the risks of thyroid disease in study participants were about the same regardless of the radiation dose they received and that if there was an increased risk of thyroid disease, it was too small to observe.

When the draft of this report was released in 1999, it created quite a storm in the scientific community. Critics of the study claimed it was fundamentally flawed because the dose estimates provided by the government were greatly underestimated and because researchers relied on people remembering the details of their diets and lifestyles more than forty years earlier. Unfortunately, the science does not exist today that can definitely determine whether individual cancers were caused by radiation exposure in the 1940s and '50s. There is no method to confirm the cause of most cancers.

* * * * *

Winning the lawsuit won't really fix anything. My father will still be in a wheelchair and our neighbors will still be dead. The only thing that winning the lawsuit would accomplish, in my opinion, would be to prevent something similar from happening again.

I don't care about the thyroid study. A friend who has worked in public health for many years has a poster on her wall that reads, "The enemy of public health is definitive science." Science can provide useful information; too often we let that information be the last word.

My parents live in Spokane now. Dad's wheelchair wouldn't work on the gravel driveway, and the winter country roads made them nervous. My sisters and I urged them to buy a condo in Spokane, but they love renting. After fifty-two years of maintaining a large house, a giant barn, and four other outbuildings, they are happy to let maintenance be someone else's problem.

It was hard for them to move though, for many reasons, including the giant question of what to do with the farm. None of the daughters wants to farm, a profession that has become less and less lucrative over the years. The most we ever got for our wheat, twenty years ago, was about $6.50 a bushel. Today a bushel of wheat goes for $2.40.

Mom and Dad solved their "what to do with the farm" problem by giving it to their daughters and letting us worry about it. That decision made sense. For over a hundred years the farm has been passed down for the next generation to worry about. Now it's our turn.

My sisters and I aren't sure how best to keep the farm, but we are certain of our determination to preserve it and to somehow reinvent its important place in our family legacy. All of us, including the grandchildren, recognize what that piece of land means to us as a family. "It gives you something to hold on to," as my mother puts it.

With wheat prices as low as they are, we sometimes consider restoring the land to its prefarming days of bunchgrass and pine trees. The Creek Place has always been like that—imagine if the whole farm were! What a radical move it would be, in the world's richest wheat-farming country, just to turn the land back to the grass and trees that used to grow there. Our great-grandparents had to plow it to keep it, in a way of life that was essential to who they were. I wonder if our legacy could be returning the land to what it was, if it is possible to come close to that.

The half life of the iodine-131 released over our fields was eight days, meaning that after eight days, half the iodine-131 became stable xenon-131, which is not radioactive. The radioactivity in the remaining iodine-131 was reduced exponentially by halves until it was, presumably, gone. The half life of plutonium-239, another radionuclide released over us, is 24,000 years.

I find it hard to imagine finding the marriage between half lives and bunchgrass and restoring what we had. But maybe we can. Or at least we can pretend that the plutonium-239 and the iodine-129, with its half life of 16 million years, aren't lingering out there somewhere. Maybe with the wheat fields gone we could raise Appaloosas, and, who knows, maybe the curlews would come back, nesting in the dry pockets of bunchgrass up from the creek, their long curved beaks digging down into the loose soil, rooting out the meadow grasshoppers and ground beetles that have always lived there.

Atomic Farmgirl

by
TERI HEIN

For Discussion

1. How does Hein present the realities of farm life? Is this lifestyle more effectively revealed through detailed accounts of the actual process or through the people whose lives revolve around the vocation?

2. Hein writes: "Wheat is our thing, and a thousand acres of it swaying in the breeze is, for us in the Palouse, about the most beautiful thing on Earth. We put pictures of wheat on our Grain Growers Association calendars and write poems about it when we go off to college" (p. 55). What does this book say about the connection that people develop with the land? Do you think the connection Hein writes about is unique because the land has been passed down from generation to generation? What underlying themes of "home" does she present that can be applied universally?

3. What devices does Hein use to develop and emphasize the different personalities in her community without confusing the reader with an overabundance of names? How does this method of storytelling illustrate the interconnectedness of the community?

4. Hein creates a vivid portrait of the early immigrants to the West. How does she characterize their lifestyle? How has it changed over the generations?

5. What does *Atomic Farmgirl* reveal about family dynamics? Which relationships does Hein explore at the greatest length? How has her relationship with her family shaped her life?

6. In what ways does Hein's experience with Rockette illustrate the relationships people develop with their animals? How does Rockette's death contrast with the other deaths Teri experienced during her youth?

7. Of her mother, Hein writes: "She was the only woman we knew who had been to college, which meant she knew loads of impractical things, such as the names of rivers in Africa and how to explain photosynthesis to us. When she referred to a book she was reading, she often said the author's name as if it were part of the title. 'I'm reading Hemingway's *For Whom the Bell Tolls*,' or 'I just finished *To the Lighthouse* by Virgina Woolf,' she would say to us, her daughters, who didn't ever care about these things until much later" (p. 5). How does Hein's mother defy the traditional conventions associated with the farm wife? What effect did her unconventionalness have on Hein and her sisters?

8. As present as the physical deaths of individuals in the community is the idea of a dying lifestyle. How would you characterize the elements that are dying? What effect does this process have on the community as a whole?

9. In what ways does Hein show how close-knit her community was? Does this closeness ever border on becoming an intrusion on people's privacy? What are the advantages and disadvantages to living in a community like this?

10. Throughout her narrative, Hein draws upon the Native American history associated with the land. How is this history woven into her story? What effect does this have on the narrative?

11. Although the book is nonfiction, what aspects of the author's style contribute to its reading much like fiction? What effect would the stories have if they were told in a more formal manner?

12. As an adult Hein has gone on to lead an adventurous life and move to a city. Based on what you know about her childhood, why is this both surprising and predictable?

13. Hein draws a contrast between the development of her own small community and that of Richland, a town that grew because of the creation of the Hanford nuclear plant. In what ways do planned communities differ culturally from those that develop naturally?

14. During the Cold War the government, in the name of security, secretly released radioactive materials over Washington State, which many believe had enormous health consequences for the people living there. How should we think about this government action, especially now, with our concerns about national security?

About the Author

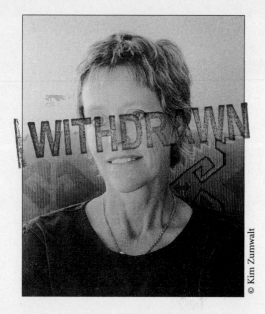

© Kim Zumwalt

Teri Hein grew up on a wheat farm originally homesteaded by her great-grandparents in eastern Washington. In the years since she left home for college, she has led an adventurous life—teaching abroad, rafting the Grand Canyon, traveling to northwestern Pakistan, doing research in the Amazon jungle, and hiking above the Arctic Circle. She received a master's degree in education from the University of Washington and has been granted awards for her teaching, as well as a Fulbright scholarship. Currently Hein lives in Seattle and teaches children who are undergoing cancer treatment at the Fred Hutchinson Cancer Research Center. *Atomic Farmgirl* is her first book.

MARINER BOOKS / HOUGHTON MIFFLIN COMPANY
For information about other Mariner reader's guides, please visit our Web site: www.houghtonmifflinbooks.com.